The Open University
Technology
A Second Level Course

Units 11–13

The Open University Press

DESIGN

PROCESSES AND PRODUCTS

BLOCK 4
CARS

FORMS AND
FUTURES

Prepared for the Course Team
by David Walker
with contributions from
Stephen Brown, Nigel Cross,
Tony Curtis, Mohammed Dorgham,
Paul Gardiner and Alex Godden

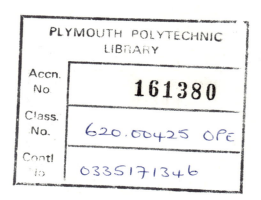
The Open University, Walton Hall, Milton Keynes, MK7 6AA

First published 1983

Designed by the Graphic Design Group of the Open University

Filmset in Helvetica by Filmtype Services Limited and Printed in England by Pindar Print Limited, Scarborough, North Yorkshire

ISBN 0 335 17134 6 ✓

This text forms part of an Open University course. A complete list of units is given at the end of this text

For general availability of supporting material referred to in this text please write to:
Open University Educational Enterprises Limited, 12 Cofferidge Close, Stony Stratford, Milton Keynes, MK11 1BY, Great Britain

Further information on Open University courses may be obtained from: The Admissions Office, The Open University, PO Box 48, Walton Hall, Milton Keynes, MK7 6AB

1.1

CONTENTS

PART THREE

STUDY GUIDE

General topic

The topic of this block is design in the motor car industry, but more than that it leads you towards making conjectures about future car designs and future vehicle technology, hence the subtitle 'Forms and futures'.

The block moves from a historical description and an outline of the basic engineering of the present-day car to the central issue of design for energy efficiency and finally to the wider contextual problems generated by the motor car.

Viewpoint

The block is largely written from a point of view within the motor manufacturing industry, of designers and engineers who are trying to make complex pieces of machinery. However, Part Three of the block steps beyond the industry to discuss choices at a strategic level. The viewpoint shifts from inside the industry to outside. Thus the later sections are more speculative and perhaps more critical than an insider's viewpoint would permit.

Aims

1   To demonstrate the interrelationship of technical and human factors in design.
2   To exemplify the complexity of the design process for a mass-produced object.
3   To indicate technological trends in car design.
4   To illustrate the wider resource and environmental issues that form the context of design.
5   To facilitate well-informed conjectures about the future of car design.

These aims are met, roughly, in the order given above. Part One of the block addresses aims 1 and 2. Part Two addresses aim 2 also and aim 3. Part Three addresses aim 4, while aim 5 pervades material in all parts of the block.

A checklist of objectives is given at the end of this block.

What you have to do

As in earlier blocks of *Design: Processes and Products*, each section here represents something like 2–3 hours' work for you. Some sections require an active involvement. You need your manikins and *Humanscale* for section 3 'Human factors'. Also a calculator would be useful for the later sections. In general you should keep your Workbook by you when studying the block, and use it to make notes as you go along. This will be particularly helpful in building up your ideas for the assignment.

This block divides into three major parts. You might like to think of it as three conventional units bound together as one.

Part One is historical in character. It provides a brief chronology of the car as an evolving artefact and some information about the basic engineering of motor vehicles. I start from a description of the way an internal-combustion engine works, and move on to problems of transmission and car body construction. Not *all* the technical details of cars are covered. I have omitted any explanation of suspension, steering, electrical system, engine accessories; rather I have concentrated on those elements that have a direct effect on fuel efficiency.

So towards the end of Part One you should have a rough grasp of the historical development of the car from the nineteenth century to the present day; you should have a working knowledge of the engineering and manufacturing problems. Then you are in a position to think diagrammatically about the form and layout of the car and about how adequately (or not) it meets human requirements. You are given an ergonomic exercise to explore this idea further.

The tutorial television programme occurs at the start of your study of this block. It sets up some ideas for the assignment and attempts to give you a general orientation to the material of the block. It has much the same purpose as 'Learning from experience' had in relation to Block 2 *Bicycles*, that is, to act as an early stimulus for your assignment.

Audiovision Package 7 'Car designers talking' should be used towards the end of Part One of the block (see the Study Chart). It relates to section 2 'The modern car', and gives you the opportunity to hear designers talking about recent developments in car design.

Part Two is narrower in its intentions. Here I describe the process by which cars are designed. There is some discussion of the management of design. I then concentrate on design for maximum fuel economy. Energy efficiency is an important theme of *Design: Processes and Products*, which you have encountered already in Block 3 *Houses*. Finally in Part Two I look more closely at one aspect of fuel-efficient design, aerodynamics.

You should try to evaluate the various means of fuel conservation both in terms of their probability and their desirability.

The television programme 'The shape of cars to come' focuses on the topic of aerodynamics. It illustrates some of the technical and historical points made in the text, but beyond that it illustrates how the design of body shapes is determined by marketing considerations as well as by engineering principles. In this way it acts as a bridge to the wider problems of car design.

Part Three steps beyond the car industry into resource and environmental concerns. The nature of our exploration becomes speculative. Two important issues are covered at length: electric power and alternative fuels. These two topics have been chosen not because they necessarily represent a plausible future direction for the motor car, but because they are intended to prompt the question, 'What if . . . ?' You may choose other topics to review in the same way, and to make those the starting point of your own hypothesis.

The final sections of the block offer you a summary of the topics and factors you might like to focus upon. I also offer you some general advice about 'Futures', about making conjectures, forecasts and extrapolations.

In this way you are brought to the point where you can undertake the assignment, in which your conjectures are informed by the descriptions of present circumstances and methods, and reviews of future probabilities.

When studying this block it might help you to keep certain key words and phrases in mind:

Part One (sections 1–3)   – *history, engineering, ergonomics*;

Part Two (sections 4–6)   – *design process, management, energy*;

Part Three (section 7–10) – *alternatives, social and environmental effects*.

It may also help you to think of the three major parts in terms of past, present and future. In detail you will find the narrative does not stick strictly to that chronological division, but the block starts at the earliest beginnings of the motor car and ends with long-term speculations.

There is no absolute necessity for you to study the block in this linear order. In fact I recommend a rather different procedure as preparation: read the assignment in the Supplementary Material to develop a sense of where the

STUDY CHART

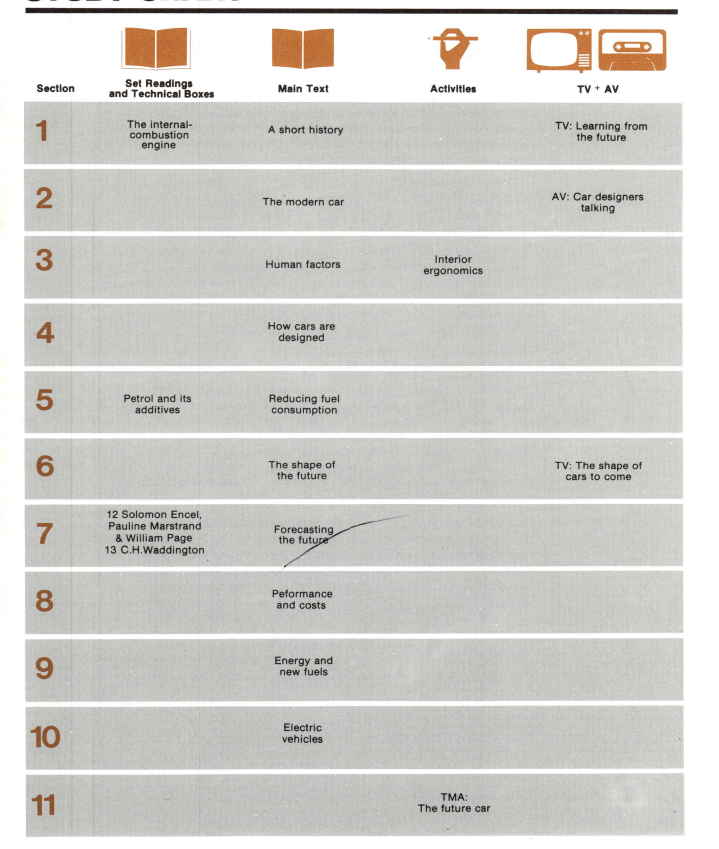

Section	Set Readings and Technical Boxes	Main Text	Activities	TV + AV
1	The internal-combustion engine	A short history		TV: Learning from the future
2		The modern car		AV: Car designers talking
3		Human factors	Interior ergonomics	
4		How cars are designed		
5	Petrol and its additives	Reducing fuel consumption		
6		The shape of the future		TV: The shape of cars to come
7	12 Solomon Encel, Pauline Marstrand & William Page 13 C.H. Waddington	Forecasting the future		
8		Peformance and costs		
9		Energy and new fuels		
10		Electric vehicles		
11			TMA: The future car	

block is leading; skip through the whole block quickly so you have an idea of its scope and the emphasis section to section; then perhaps go to section 7 'Forecasting the future' in order to think a little about your overall strategy for the assignment; perhaps reinforce that by looking at the more speculative examples in the final sections; then, hopefully, you are in a purposeful frame of mind and able to take on the block in its linear order.

You may find the basic engineering described in Part One very familiar. It covers the mechanics of conventional cars up to the recent past. The details of Part Two may be less familiar to you. Here the focus is upon current research in the car industry, upon improvements in design that have only just, or not quite, come into production. As you will see, the car industry is on the threshold of many exciting new technologies.

The technical details of Part Three will probably be more unfamiliar to you. You are asked to step back and look at the future of the car and its effects more dispassionately, to make a balance sheet so to speak. Finally, the block culminates in a review of two major technological developments that may affect long-term trends.

In your assignment you are asked to make some kind of speculation about the car. There is no way of assessing these speculations as right or wrong (unless you are willing to wait twenty years for your tutor's response), but your assignment can be assessed as well-informed, well-argued and well-structured. The material of the block is intended to help you achieve that end.

The scope of this block

The motor car has its own specialized technical press. Each month many thousands of words are devoted to recent developments in the car world. It might seem that nothing new could be said about the car. It might appear that, if you are a keen reader of motoring magazines, this block can contribute nothing further to your understanding. But the motoring press, as a whole, operates within narrow limits: it deals mainly with the latest artefacts, it reviews them as consumer products, it tends not to be too critical, and it assumes a high level of technical familiarity or, at least, a familiarity with the jargon.

This block approaches the car in a different way. I do not assume you have a great technical knowledge of the car, so the block gives you an outline of the basic principles of the engineering of cars. It gives you a historical review from the earliest beginnings to the present day. It tells you something of the design, management and manufacturing process. It discusses the kind of issues that preoccupy designers and engineers at the present time. Finally, it raises some critical doubts about the motor car and its costs.

PART ONE

1

A SHORT HISTORY

1.1 Introduction

This section is in part a description of the basic mechanics of cars and in part a history. One of the functions of history is to give you an apprehension of what might happen next. This short history may not be prophetic, but I hope it will convey a sense of direction and may help you to come to a view of what is probable as a future for the car industry and for car design. But more than that it will give you an understanding of the range of possibilities. Some of these possibilities have been tried and dropped. Others, more interestingly, have yet to be tried. They await new technologies and new attitudes.

You may feel that the kind of motor car now on the road is inevitable, the logical outcome of rational forces, or at least the best possible balance of conflicting requirements. But I should like to warn you against that attitude, in the sense that, for the designers and engineers close to the problems, working in the present, the eventual outcome of their efforts is not at all clear. For example, before 1959 no one was able to predict the huge influence that small front-engine cars would have on the industry in the 1960s and 1970s.

In retrospect, a succession of technological developments tends to look tidier, more ordered, rational and predictable than such developments looked at the time of their unfolding. You will be asked at the end of this block to make your own kind of predictions, or broad conjectures. This is an intrinsically difficult task. Experienced car designers now find it difficult to be very certain about the future of the motor car. But it is worth remembering that in the past the feeling of uncertainty was exactly the same. Many possible pathways were available at any given time. So your task is not so much to choose the *correct* pathway, as to say *what if* this particular pathway were followed. This attitude, 'What if . . . ?', can apply in most sections of the block and we can apply it to historical material in this first section. Let me give you two short preliminary examples of what I mean by that.

Steam car

In 1784 James Watt had devised a steam-powered carriage, but the first successfully operating steam car appeared in 1801. This was the design of Richard Trevithick, a thirty-year-old Cornishman (Figure 1). Over the next two decades England led the way in the development of the steam carriage.

However, there were many whose livelihoods were in some way connected with horses, for example, trustees of toll roads, stage-coach proprietors, innkeepers and so on. To prevent the development of the steam carriage, punitive bills were hurried through Parliament. As a result excessively heavy tolls were imposed on steam vehicles. For example, on the road between Liverpool and Prescot a loaded stage coach paid only 20p toll while a steam carriage had to pay £2.40. In 1831 a select committee of Parliament recommended that these heavy tolls should be lifted, but no action was taken. As a consequence, engineers and entrepreneurs diverted their attention to forming a new system in competition with the roads: the railways.

This raises the following strange speculations: *What if* there had been no impediment to these early steam carriages? They could have developed into a more sophisticated steam car, become widely used on the road, and the railway system would have been inhibited, or even pre-empted by a road-based network of tracks.

Figure 1 Richard Trevithick's steam coach

This vehicle was successfully demonstrated in London in 1803

Figure 2 A 'locomobile', United States of America, around 1900

A popular steam car that was produced in large numbers at the beginning of the twentieth century

The duration of life of such British steam cars, at the least, may have been equivalent to that of the American 'Locomobile', shown in Figure 2. This steam car was very popular at the turn of the century. Up to 1901 it accounted for 4000 vehicles, or *half* American production. It even has the legitimate claim to be the first mass-produced car.

The continuing suspicion of new forms of vehicle in Britain prompted the speed restriction known as the Red Flag Act of 1865. This Act made speeds above 4 m.p.h. (6 km/h) illegal in the country and above 2 m.p.h. (3 km/h) illegal in the town. Road locomotives were subject to a further indignity in town. They were preceded by a man carrying a red flag. This Act was revised in 1896 and road vehicles were permitted to travel at up to 12 m.p.h. in the open country and 5 m.p.h. in the town.

11

Model A Ford

My second example concerns a properly developed motor car of a later date.

The impact of Henry Ford and his first mass-produced car, the Model T, is well known. I shall deal with it a little later. What is less obvious is the *ethic* that informed the Model T and its successor the Model A (not to be confused with the first Ford Model A, a 'quadricycle' as Ford called it).

The Model A was launched in December 1927, but by 1932 production had ceased. An example is shown in Figure 3. It could not be called an unsuccessful car commercially (having sold almost five million), but the Model A was overtaken by its competitors, and a different set of principles. Henry Ford's policy with the Model T and Model A was to mass-produce a *durable, basic quality vehicle at low price*, and gradually to improve the product with a minimum of radical changes. However, General Motors had introduced a new ethic. Firstly, they introduced the concept of an annual change of model, by updating or 're-skinning' the basic engineering. Secondly, General Motors, as early as 1923, offered a carefully *structured range* of cars, which enabled customers to progress steadily up the ladder of prices while remaining loyal to one brand, thus:

from a Chevrolet at	$525–775	(1923 prices)
through a Pontiac at	$825	
an Oldsmobile at	$875–1115	
an Oakland at	$975–1295	
and a Buick at	$1125–1995	
to a Cadillac at	$2995–4485	

This structuring of the price bands of cars is known as 'Sloanism', after Alfred Sloan, who was then Chairman of General Motors. (Information from Burgess Wise, 1982.)

The progress of the car industry from the 1920s very largely followed the ideas of Sloan rather than those of Ford. It became a self-consciously market-oriented industry, rather than an industry devoted to cheap basic long-life products. There are exceptions to this general pattern, like the Volkswagen 'Beetle', but it is tempting to speculate about the state of the industry today, and over the intervening sixty years, if it had followed the first principles of Henry Ford rather than of Sloan.

Figure 3 Ford 1928 Model A Special Coupé

By 1932 the Model A had reached sales of over five million

What if these examples of the steam carriage and the durable basic motor car are put together?

If I put these two examples together, they predicate a widespread, durable, basic, long-life steam car or bus as the major form of transport. However, the internal-combustion engine quickly came to dominate the propulsion of motor vehicles.

Let's look at the chronological development of early cars and their engines.

1.2 Early history

At the simplest level a modern car consists of four wheels, an engine and power-train to these wheels, accommodation for driver, passengers and luggage, and a covering body shell. Other elements, such as chassis, suspension and electrical system, are important in the design of cars, but are less important in their impact on fuel efficiency. These opening sections will review some of the possible arrangements for the basic elements: engine and transmission. Wheeled vehicles powered by internal-combustion engines were the outcome of three existing lines of technological development: steam engines, carriages and bicycles.

Therefore it is hardly surprising that the very first automobiles (self-propelling vehicles) looked like the mongrel offspring of the three technologies.

The inventors of this period faced three main problems. The first was the development of an efficient, reliable engine with enough power to drive the vehicle and its occupants. The second was the problem of transmission and the matching of the engine; the third problem was the general configuration of a wheeled vehicle. In this section I shall concentrate mainly on the development of the engine.

Figure 4 De Dion–Bouton Voiture Steamer 1885, with front-wheel drive and rear-wheel steering

The internal-combustion engine

Principles

An engine converts the chemical energy in the material of the fuel into mechanical energy. In an internal-combustion engine the fuel is usually petrol or diesel oil.

In order to unlock the chemical energy in petrol it has to be burnt in the presence of oxygen in a controlled explosion. In a petrol engine, a mixture of petrol vapour and air is fed into the combustion chambers, then compressed and ignited. After explosive combustion the burnt gases are expelled and the process begins again. The force of the expanding gases in the combustion chamber pushes a piston in a power stroke along a cylinder. A connecting rod from the piston to a cranked shaft (crankshaft) translates linear motion into rotational motion.

Four stroke. The four-stroke internal-combustion engine operates in a cycle of (1) induction (down), (2) compression (up), (3) power (down) and (4) exhaust (up) (see Figure 5).

Two stroke. The two-stroke engine is simpler in operation in that the compression (up) stroke introduces fuel *below* the piston; the power (down) stroke simultaneously exhausts gases and *transfers* fuel to the top of the cylinder. Two-stroke engines are widely used in motor cycles, but have lost favour in cars because their exhaust gases contain high proportions of unburnt and toxic material.

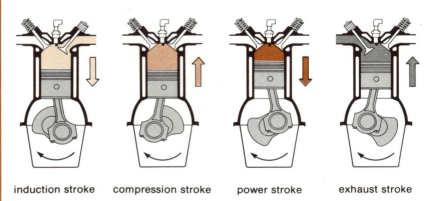

induction stroke   compression stroke   power stroke   exhaust stroke

Figure 5 Operating principles of a four-stroke engine

Numbers of cylinders

A car engine may work on any reasonable number of propelling pistons (or cylinders if I use the normal term). *Even* numbers of paired cylinders coupled to the main drive shaft have been shown generally to produce a balanced smoother running. Thus it is common to find engines with 2, 4, 6 and even 8 cylinders, but more rare to find 1, 3, 5 or 7 cylinders.

Arrangement of cylinders

The arrangement of cylinders about the crankshaft is as various as can be imagined. Examples include the early V-twin of Daimler and Maybach (see Figure 12) to the later well-known Ford V8 of 1953. The Volkswagen engine was arranged as a flat four cylinders, that is, as horizontally opposed double pairs. In Germany this arrangement was known as a 'boxermotor' (Figure 6). The Citroën 2CV engine has one pair of horizontally opposed cylinders.

Figure 6 Four-cylinder horizontally opposed engine, or 'boxermotor'

This engine was developed by Reimspiess in 1937 and later used in the Volkswagen 'Beetle'

The arrangement of the cylinders is very largely determined by how the engine designer wishes to apply power to the driving shaft.

Figure 7 shows various arrangements of pistons and crankshaft. The engine designer can manipulate the number and geometry of the cylinders, the point at which the power strokes are applied to the crankshaft, the time sequence of the pistons and the shape of the crankshaft itself.

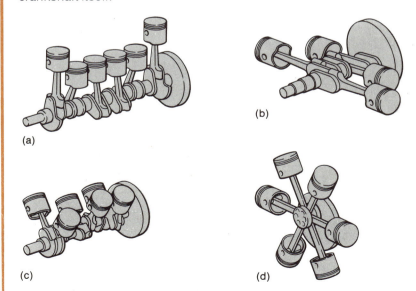

(a)

(b)

(c)

(d)

Figure 7 Various arrangements of pistons and crankshaft

(a)   In-line six-cylinder engine (b)   Horizontally opposed engine
(c)   V-engine (d)   Radial engine

Power

Once the engine is in motion the tight fit of the pistons within the cylinders sucks the fuel mixture into the compression chambers. The flow of fuel is controlled by a valve (or throttle) activated by the accelerator pedal. When the pedal is depressed the throttle opens and an increased flow of fuel mixture is allowed.

The admission of air and fuel into the cylinders is controlled by further valves operated by a camshaft, which in turn is driven by the crankshaft. Thus the opening and closing of the valves is mechanically synchronized to the movement of the pistons.

The most efficient burning of the fuel mixture is achieved when it has a swirling motion imparted to it as it enters the combustion chamber. This turbulence is to ensure that the air and petrol vapour mix uniformly, otherwise pockets of rich mixture would form giving unburnt exhaust gases and loss of power. For a given maximum engine speed and compression ratio, the power that can be developed by an engine is directly proportional to the volume in the cylinders available to the explosive petrol mixture (known as the cubic capacity). If one wished to increase the power of an engine then the simplest method is to increase the cubic capacity. There are three ways to do this (Figure 8):

1   increase the stroke of the piston – length,

2   increase the cylinder bore – diameter,

3   increase the number of cylinders – number.

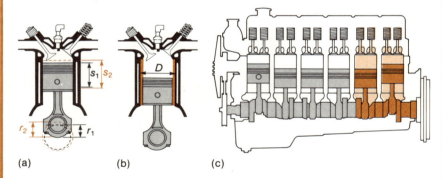

(a)   (b)   (c)

Figure 8 Three means of increasing the cubic capacity of an engine
(a)   Increasing the stroke (b) Increasing the cylinder diameter (bore)
(c) Increasing the number of cylinders

Early motor cars were small, light and simple, out of technical *necessity* rather than desirability. The first engines were heavy and not very powerful. A slow-speed engine prior to 1884 developed 1 horsepower (h.p.) for every 300 lb of engine weight. However, in that year Gottlieb Daimler constructed a petrol engine that developed 1 h.p. for every 90 lb of engine weight. These engines prior to 1900 had a cubic capacity of about a litre and developed typically 2–4 h.p. The petrol engine therefore fulfilled the requirement of a high power-to-weight ratio. (Caunter, 1970, pp. 1–2.) In 1885 Daimler and his collaborator Maybach mounted their engine into a wooden bicycle frame, thereby inventing the first motor cycle.

Slightly earlier in that year Benz successfully combined a somewhat slower, less efficient two-stroke engine with a tricycle to produce the first three-wheeled 'car', Figure 10.

At Paris late in 1885 Daimler saw Benz's three-wheeled car and quickly went on to produce his four-wheeled car the following year. Daimler and Benz then remained very bitter rivals for the rest of their lives, although ironically their two firms were merged as Mercedes–Benz in 1926.

Figure 9 Model of Niklaus Otto's original production engine of 1872

Although it was very noisy and inefficient and could never have powered a road vehicle, Otto's was the first internal-combustion engine to use the now universal four-stroke cycle

Figure 10 Karl Benz's pioneering three-wheeler of 1885

Figure 11 Competitors in the Paris–Rouen Trials of 1894

The winning De Dion steamer was no. 4. No. 24 is noteworthy for its steering wheel, the only example in the event

In the ten years after 1886 many imitators, entrepreneurs and improvers followed on from the initial co-inventions of Daimler and Benz. The various forms of engine were tested out in prototypes. The Paris to Rouen Trials of 1894 present us with a wide range of car designs, some of which are shown in Figure 11, and give us a picture of the state of vehicle technology at the end of the nineteenth century.

The Trials were open races for propellors of all kinds. Entries included such improbable means as gravity, hydraulic mechanisms, systems of pendulums and 'self-acting' mechanisms (but none of these materialized). The 102 starters included steam, electric, internal combustion, mainly as three- or four-wheeled vehicles. Of that 102 only fifteen managed to finish. A De Dion steamer was the winner, but the remaining fourteen finishers were powered by Daimler engines, some in Daimler cars and some in cars by other designers.

While steam was the outright winner, clearly the Daimler internal-combustion engine emerged as the best design when compared to competing electric and steam engines. Internal-combustion engines rapidly became the technological leaders during the 1890s because they offered increasingly better power-to-weight ratios. Higher and higher speeds of rotation at lower and lower levels of fuel consumption made diesel or petrol engines much better choices for powering wheeled vehicles. For the same reasons, von Zeppelin began using internal-combustion engines for his rigid airships in the 1890s, as did the Wright brothers for their flying machines in the 1900s. By the late 1890s cars powered by internal-combustion engines appeared in a number of different countries.

Two further unrelated events of 1896 are of significance. In Britain the Red Flag Law for powered road vehicles was repealed, thus removing one obstacle to experimentation and innovation in vehicle design. In the American state of Michigan, Henry Ford built his first motor car. Although this was what

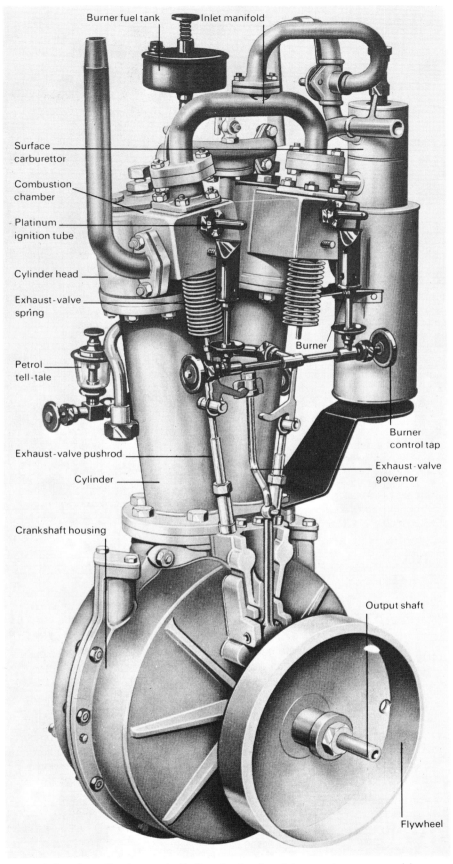

Burner fuel tank
Inlet manifold
Surface carburettor
Combustion chamber
Platinum ignition tube
Cylinder head
Exhaust-valve spring
Petrol tell-tale
Exhaust-valve pushrod
Cylinder
Crankshaft housing
Burner
Burner control tap
Exhaust-valve governor
Output shaft
Flywheel

Figure 12 Daimler–Maybach engine of 1889

can only be described as an imitative design he did go on to a reasonable commercial success.

Between 1896 and 1908, when Ford began to mass-produce his Model T, a number of important technological thresholds were crossed.

Look at Figure 13. This shows how a cluster of patents came together over 60 years and coalesced as a new type of artefact.

19

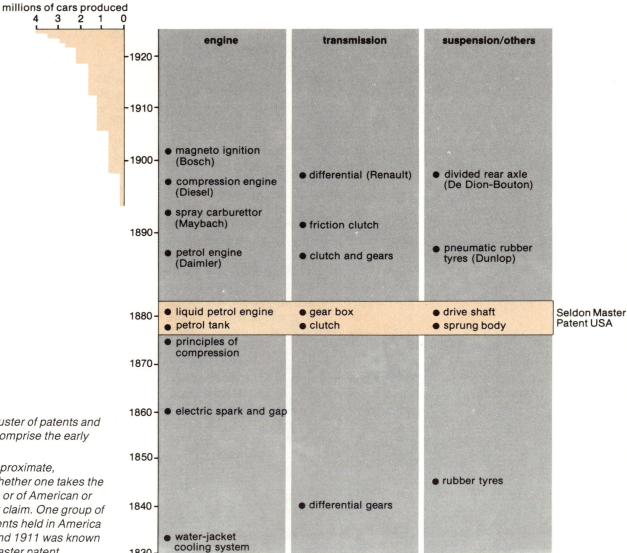

Figure 13 The cluster of patents and inventions that comprise the early motor car

The dates are approximate, depending on whether one takes the date of invention or of American or Euorpean patent claim. One group of six separate patents held in America between 1879 and 1911 was known as the Selden master patent

One patent, first proposed by George B. Selden in 1879 and granted in 1895, grouped together the six major ideas that comprise the automobile. This is sometimes known as the Selden 'master' patent. American manufacturers paid Selden a royalty of $1\frac{1}{4}$ per cent on the retail price of all cars sold. Henry Ford successfully challenged the validity of the Selden patent in 1911, and the Selden claim was not an inhibition on the growth of the motor car in the United States.

The number of cars produced is shown on the top left of Figure 13: from 300 in 1900 to over *4 million* in 1925 in the United States alone. During this period the motor car changed from being an entertaining toy for the wealthy to a utilitarian means of transport, widely available.

SAQ 1

Sketch the main arrangements of cylinders in an internal-combustion engine.

SAQ 2

Why did the internal-combustion engine become more important than its rivals in steam and electricity?

SAQ 3

What is the historical significance of the 1890s for the motor car?

1.3 Batch production to mass-production

The first cars were unique, one-off artefacts. They were craft built by the inventor and perhaps a small team. As the industry grew the methods of craft building changed to batch production of short runs, with some standardization of components. Before the First World War, however, mass-production had begun in America. The, then, novel and elaborate methods of manufacturing had a profound influence not only on the growth of the industry but upon the way cars were designed henceforth. Let's look at the early development of the manufacturing processes.

Initially the car was viewed simply as a powered carriage. To support the engine and to locate it accurately in relation to the driven wheels required a mounting frame. On the earliest machines the frame was quite light (e.g. 1899 De Dion–Bouton), but as the engines became larger, this 'chassis' became a substantial pair of girders, sometimes forged, otherwise fabricated from channel section, which provided the structural frame for the rest of the functional elements: the engine, the wheels, suspension and steering, and the passenger coachwork. Compare the De Dion–Bouton chassis in Figure 14 with the Morris chassis of 1929 in Figure 15.

It is understandable that the inventors, designers and manufacturers of these cars (often the same person) saw the chassis, wheels and engine as a powered trolley, and the body or cab as an addition to it.

You can consult photographs of early batch production layouts and see how the cars were assembled. In 'batch' production ten or twelve units were assembled at a time, the components being brought to each in turn. Starting with the chassis on a couple of trestles, the axles and wheels were attached, then the engine and steering, followed by various mechanical components, and lastly the coachwork decked the car out.

Figure 14 Chassis of a De Dion–Bouton of 1899, which is of light tubular construction

Figure 15 A Morris chassis of 1929, with substantial channel sections bolted together

21

Figure 16 Batch production methods at the Iris Car Works, Willesden, in 1907

Chassis are laid out on wooden benches in a 'hospital ward' arrangement. There is no indication of any kind of power tool used in assembly. The Iris went out of production at the beginning of the First World War

If you were a practical do-it-yourself mechanic, you would adopt the same sequence if you were thinking of making only *one* car. For instance, apprentices at a car works today designing and building a competition go-cart follow much the same route. If things don't quite fit or require modifications, they can be done on the spot by a competent fitter. Components do not have to be very precise or made with closely specified tolerances, but a second car would not be quite like the first. And so it was with early cars.

Batch production is acceptable only as long as you don't have *too* many articles to produce. If the workshop can hold about two days' output, there can be a daily clearing of half the work and the laying out of the next day's chassis and some local stocking of components. There might be a need for doors at each end of the workshop and additional space to facilitate movement. Usually this method entailed a lot of walking between units by the fitters and assemblers, and there would also be a lot of fetching and carrying of components from the stores.

With several hundred car producers operating in America alone in the first decade of this century (estimates are that in 1911 there were 270 manufacturers producing 400 models) sooner or later one was to discover the natural advantage of using the *wheeled* chassis as an assembly bed in a progressive line rather than keeping to the hospital ward layout of static parallel 'beds'. It took the dynamism of Henry Ford, who wanted to be the builder of the first popular, or 'people's' car, to see this assembly line progression applied to the *whole* car production.

Thus the car lent itself readily to mass-production because it had wheels.

The improvement in manufacturing tolerances and the move towards standardization were a crucial step towards mass-production. In 1908 three Cadillacs were shipped to England, were completely dismantled and the parts were mixed. Some randomly chosen new parts were substituted to ensure that the three originals were not specially constructed for such interchange. Under the supervision of RAC officials acting as judges, parts were chosen at random, and the three cars reassembled using only four tool types: screwdrivers, spanners, hammers and pincers. The assembled cars were then taken to Brooklands and driven for 500 miles. The Cadillac Company was awarded the Dewar Trophy for the most significant demonstration of motor car technical progress in that year. This dramatic demonstration was a repetition of a similar one done by Eli Whitney with muskets in 1798.

Figure 17 Ford started producing cars in Britain in 1911, when this factory at Trafford Park, Manchester, was opened. From then right through until 1932 (with one exception) the factory built only right-hand-drive versions of two American models, Models T and A. Ford was one of the first people to introduce a moving assembly line in America, and by the time this picture was taken at Trafford Park in 1914, production-line assembly here was also automated.

The moving production line is on the right and you can clearly see the means of supplying the wheels to the line. On the left are vast stores of engines, while in the background at the centre the line of completed bodies can be seen coming in at right angles to join the chassis line. In most other factories, of course, the chassis had to be taken manually to the body shop for this operation. Unlike the Highland Park plant in America, Trafford park did not have the famous 'body drop', bringing coachwork down from above to avoid the problem of lifting it on to the chassis, as would be the case here.

By 1929 Ford was making one vehicle every 2.8 minutes of the working day at Trafford Park

The standardization of components leads logically to complete assemblies of standard artefacts. Henry Ford had a vision of motor cars that, in his own words, would be 'all alike . . . just like one pin is like another pin when it comes from a pin factory, or one match is like another match when it comes from a match factory'. This vision was to be realized only when the standard parts could be assembled in a standard system of extraordinary efficiency.

When Ford conceived of mass-produced, low-cost Model Ts he thought out the assembly system simultaneously with the design. He devised the tributary lines that fed the main line with assembled components, flowing from the parts stores and incoming delivery areas ('goods inward bays'). He used conveyor belts, continuous chains and gravity feeds to move the parts, components and subassemblies around. Man and tools remained in one position; the work moved. Assembly followed a routinized progression, adding to the chassis until the coachwork was finally added as a single unit. By these methods production rates of assemblies increased hugely over previous methods.

In 1909 the Ford Company produced 10 607 cars at a rate of 200 per week. The following year this had almost doubled. The progress after that is indicated in Table 1.

Table 1 Ford production, 1912–14

Year	One car time	Total production	
1912	12 h 30 min	78 440	small moving subassembly lines
1913	6 h	168 220	complete moving assembly introduced
1914	1 h 33 min	248 307	

Source: Burgess Wise (1982).

By 1915 the new Ford factory at Highland Park, Detroit (Figure 18), was producing 3000 cars per day. The period of design and development for the Model T was from late in 1906 to September 1908, just under *two years*. The core of the design team consisted of *four* people: Henry Ford, Charles Sorensen (who specialized in casting iron), Joseph Galamb (a Hungarian engineer) and Childe Harold Wills (a metallurgist).

Figure 18 The Ford factory at Highland Park, Detroit, built in 1915

In Unit 1 *An Introduction to Design* I gave you a period of six years as the usual time of development of a modern car, with many hundreds of designers and engineers working on the design. A modern car is very much more complex than the Model T and the constraints and inhibitions on design are much more fierce now than they were in 1908. Yet it would be a mistake to minimize the magnitude of the Model T programme. It represented a sum of $200 000 in experimentation and $150 000 in new production machinery, a total cost of $2 201 150 at 1983 prices. The cost of the Model T tourer in 1915 was $490, about £3000 at current prices. (These current prices are based on a tenfold increase in the American price index between 1908 and 1982 and the January 1983 exchange rate of $1.59 = £1.)

During the period 1908 to 1927 over fifteen million Model Ts were produced, and the price fell by two-thirds.

The methods of Henry Ford were taken up by other manufacturers. notably William Morris and Herbert Austin in the United Kingdom and Louis Renault and André Citroën in France. In 1925 Morris was making 55 000 cars a year, not quite in the same league as Ford, although that figure represented over 40 per cent of total British production at that time.

The outcome, as you are aware, is that almost all cars now are mass-produced. Current thinking suggests that a modern car needs to sell *one million* units in order to repay the investment costs and to be profitable. Cheap cars can only be produced from a very complex system of high investment, centralized plant, standard parts and components, and moving assembly lines.

SAQ 4

What are the essential features of mass-production?

SAQ 5

What were the achievements of Henry Ford as exemplified in this section?

1.4 Television programme 'Learning from the future'

At around this point it is appropriate for you to view the first television programme in this block. This programme is a 'tutorial' programme and in many ways is similar to the earlier programme 'Learning from experience'. Both programmes show a group of students working on a project that is in essentials the same as your assignment. But in the first case the task was to *design* an improvement to a bicycle while here you are asked to *make conjectures* about the future design of cars. So, whereas in Block 2 *Bicycles* you took on the role of a designer, here you are asked to take on the role more of an analyst and forecaster.

The purpose of the television programme is to act as an early stimulus for your assignment, not so much in the sense of offering you specific ideas to follow, but rather for you to move towards a decision about which general area of car design you are going to concentrate upon.

Programme notes

The programme was made at Coventry (Lanchester) Polytechnic with first-year students on a BA honours degree course in Industrial Design. The group of students were all specializing in Transportation. Their brief for the project was approximately the same as for your assignment.

Their graphic skills may be, to your eyes, quite sophisticated. But look beyond the form in which ideas are presented to analytical skills and the way evidence for a particular line of thought is put together.

The students who appear in the programme are:

Rupert Cambray (aerodynamics), Geoffrey Bird (electric vehicles),
Kevin Rice (communal vehicles), Mark Goodall (2/4-wheeler combinations).

Their tutor is David Browne, Senior Lecturer in Industrial Design (Transportation).

The programme is presented by Nigel Cross.

Before viewing

Look back to certain elements of Block 2 *Bicycles*, particularly sections 6.2 and 9.2, which give you some sense of how to *classify* ideas (as well as how to generate them).

Look forward to the assignment for this block. Read through the Supplementary Material carefully and make an initial note of the first ideas in your Workbook. It does not matter if you feel they are naïve or uninformed ideas at this stage. At the least it will be interesting to see if your preliminary thoughts are developed in the programme by the Coventry students.

Take a brief look also at section 7 'Forecasting the future', which explains some of the principles of writing 'scenarios' for the future.

During viewing

Keep your Workbook handy for the occasional jotting to refer to when the programme is over.

It may help you to perceive the programme in three tiers:

1 The content of the students' work, their ideas.

2 The evaluation of that work for its plausibility and connection to historical and current trends.

3 The processes of argument and conjecture used by the students.

In the first case the students speak for themselves; in the second case their ideas are commented upon by their tutor and assessed against recent history; in the third case Nigel Cross makes comments and offers advice on the methods used.

After viewing

Try to summarize the main points of the programme while they are still fresh in your mind. Try to recall and note down the general lessons suggested by Nigel Cross at the end of the programme.

Read through the following general notes, and then attempt SAQ 6.

You may remember that in Block 2 *Bicycles* you were offered three levels of generality in considering the features and performance of bicycles (p.107). These three levels are useful for you now in considering cars:

1   Features of the car.

2   Types of car.

3   Alternatives to the car

After the programme, when thinking about the ideas of the students, it would help if you identify the level at which the conjectures are being made. Is idea X at the level of:

one part or feature of the motor car?

a different type of car?

a revision to the transport system?

This classification of part, whole artefact, or system should be familiar to you from your study of Block 2 *Bicycles*.

It is also important that you think about the time dimension in relation to conjectures made in the programme. Is idea X something which is realizable:

in the short term?

in the medium term?

in the long term?

These two very general ideas, the *scope* of the change to the artefact and the *time* dimension for it to come into effect, are important in thinking about your assignment and in making your own conjectures.

SAQ 6

From the range of ideas put to you in the television programme, classify them in the following matrix.

Time ----------------------------- →

Scope	Short term	Medium term	Long term
Features of cars			
Types of car			
Alternative systems			

Bear this matrix in mind as you read through the block and fill it in as you come across new information.

2
THE MODERN CAR

In this section I am going to discuss two further elements of motor cars: transmission and body shell. I have selected these not only because they are fundamental to understanding the basic engineering of cars, but because both are on the threshold of major changes. As you will see in Part Two, new designs of transmission and new forms of body shell will contribute greatly to improvements in fuel efficiency. In order for you to appreciate the basis of these improvements, it is necessary to cover, first, the more or less conventional solutions of the recent past.

This means that my technical explanation here is selective. Certain elements, such as suspension, steering, electrical system and engine accessories, have been neglected. This does not mean those things are unimportant, but in our general perspective of thinking about the car of the future and thinking about fuel efficiency, some elements are going to be more influential than others. I have chosen to concentrate on areas where radical changes are happening.

2.1 Problems of transmission

We have seen how an internal-combustion engine generates rotation at the crankshaft. The engine of a modern car develops a maximum of around 5000–8000 revolutions per minute (r.p.m.). In order to transmit the rotating force (or torque) at more typical speeds, from the crankshaft to the wheels there are considerable engineering problems to be overcome.

The main difficulties of transmission are (see Figure 19):

(a)   The speed of the engine is very much in excess of the required speed of rotation of the wheels. Even when travelling at high speeds the road wheel of a modern car is still rotating, in relative terms, quite slowly. For instance at 60 m.p.h. the rotation would be about 720 r.p.m. for a typical size wheel.

(b)   The propeller shaft traditionally runs at right angles to the rotation of the wheels. The engine and crankshaft can be re-organized to drive the wheels more directly, as in a transverse-engined, front-wheel-driven car. But Figure 19 shows a more typical in-line engine with traction to the rear wheels.

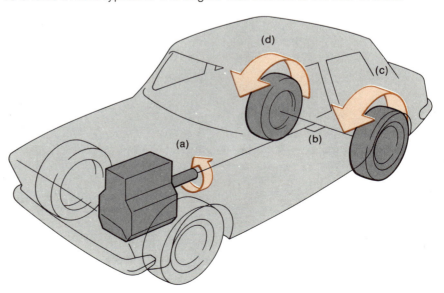

Figure 19 The problems of transmission

The rotational force (torque) generated at the engine must be transmitted to the wheels

(c)   More important still is that the power of the engine has to be utilized at various road speeds. The car is required to start from rest, to accelerate slowly or quickly, to perform well when heavily loaded or when climbing hills, and to cruise at high speeds on level roads.

When riding a bicycle perhaps you have noticed the way the gears allow you to put in the *same* effort but under different road conditions the bicycle moves differently. You pedal rapidly in a low gear to move slowly up hill. As the human engine you are working at your maximum r.p.m. to overcome the gradient. When you ride on a flat road, you can pedal quite slowly in a high gear. These principles hold true for all forms of ground vehicles.

(d)   The power of the engine has to be applied to the wheels by means of a differential, or final drive. This is made difficult, not only by the change in direction to be accommodated, but also by the fact that the wheels when cornering turn at different speeds. The outer wheel is moving on a larger turning circle, therefore it has to travel farther, and therefore has to rotate faster.

Let's look at how one of the first cars attempted to deal with these problems. You have already seen an illustration of the Karl Benz three-wheeler of 1885 in Figure 10. Figure 20 shows the transmission geometry of that car. I have shown the rotation of the various components in brown. Inspect Figure 20 and trace the propulsion from piston to rear wheels.

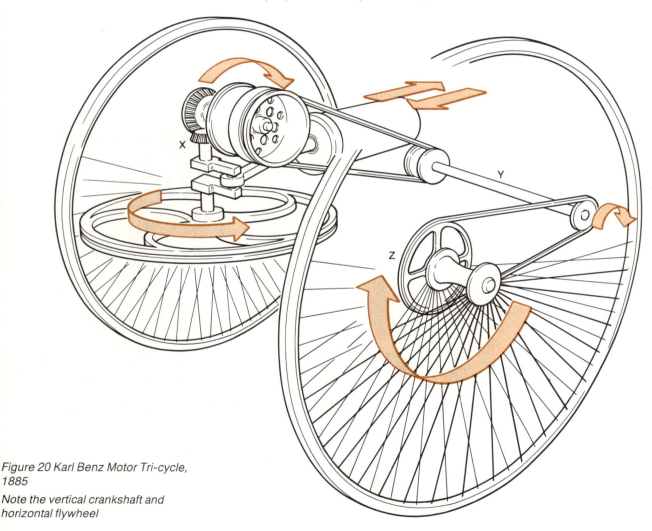

Figure 20 Karl Benz Motor Tri-cycle, 1885

Note the vertical crankshaft and horizontal flywheel

How does Benz solve the transmission problems indicated in Figure 19?

Engine speed to wheel rotation. The single cylinder is placed over the rear wheels and drives a vertical crankshaft at about 300 r.p.m. maximum. This speed has to be matched to the wheel rotation. The diameter of the wheels is about 3 ft 6 in, giving a circumference of 11 ft. If the vehicle is to move at, say, $7\frac{1}{2}$ m.p.h., the wheels must revolve once a second (22 ft per second = 15 m.p.h.) That gives a wheel speed of 60 r.p.m., from 300 r.p.m., a reduction of 5:1 from engine to wheel. This reduction occurs between X and Y, and between Y and Z. The crown wheel revolves only once for every two rotations of the crankshaft and the layshaft revolves $2\frac{1}{2}$ times for every single revolution of the wheel.

Because the engine has a single cylinder, Benz used a flywheel to maintain momentum and give smoother running. You may notice the flywheel is similar in form to that on a treadle-operated sewing machine.

Direction of propulsion. The crankshaft on the Benz is vertical. The rotational force is turned through 90° by two bevelled gears at X. This is very similar to modern solutions. The driving force is now in the right axis but in the wrong position vertically. The assemblage of pulleys, belts and chassis brings the traction to both rear wheels.

Different gears. This particular vehicle has only one gear. The belt drive could be modified to slip from one diameter drum to another beneath the driver's seat. Certainly a later Benz car used such an arrangement to give three gears, with typical road speeds of 2.6, 6 and 14 m.p.h.

Figure 21 Benz engine of 1900 in the 3 h.p. 'Comfortable'

The flywheel is now spinning vertically, with two drum sizes to the belt drive on the right

Cornering. Early cars were narrow from wheel to wheel (track), about 3 ft 9 in in this case, and their speeds were low. The differences between the two wheels when cornering was not much of a problem. This car has a divided countershaft fitted with a differential gear.

Clutch. When starting from rest in a modern car the engine is not connected to the wheels. Similarly, when gear changes are made the engine and wheels are momentarily decoupled.

Modern cars use a clutch to connect the engine power gradually to the drive-shaft.

The belt drive acts as a form of clutch. The belt is shifted into contact with the drum by a lever, which could also act as a brake on the belt. Bursts of power would be absorbed by the belt slipping, until friction begins to pull the belt around at the speed determined by the engine. This is on a large-diameter drum, so the belt never travels at the maximum engine speed. On later Benz cars, when changing gear, the belt moves from one drum to another and there would be a moment or two before it engaged with the engine.

Next let's look at the transmission of a typical modern car. Bear in mind that, although the mechanics are different and more sophisticated than that of the Benz tri-cycle, the problems that confront designers are essentially the same.

Figure 22(a) shows the transmission layout of a conventional medium-size car. Let's follow the pathway of the power train, from the engine to the wheels.

A typical in-line four-cylinder engine produces high-velocity rotation at the crankshaft. This rotation is made smooth by a flywheel rigidly connected to the crankshaft. The flywheel in turn is in contact with a driven clutch plate, by means of a spring diaphragm (Figure 22b). The two surfaces are dry and non-slip. Thus they rotate at the same speed. When the driver wishes to disengage the engine power, the clutch is depressed, a gap opens up between the diaphragm and clutch plate side of the flywheel. The engine continues to turn, but no power is transmitted.

Further down line is the gearbox, a series of interconnecting cogs on two axes. You might like to think of this as the layshaft, cogs and drums in the Benz vehicle all compressed into the same place, rather than operating distantly through belts and chains.

In a modern gear box (Figure 22c) the torque from the input shaft is transferred to the layshaft and then again to the output shaft. The two transitions give the required gearing.

If all the gears were fixed rigidly on their shafts then no movement at all could occur. So all the gear wheels on the output shaft can spin independently. The driver moves the gear lever, which selects which gear wheel is to be locked into position. This locking is achieved by sliding collars between the gear wheel and the output shaft. When one of these is slid into position along the axis of the shaft, projecting teeth (or dogs) engage with the matching projections on the gear wheel; then the output shaft, gear wheel and the collar rotate as one unit. Most manual gearboxes have one reverse gear and four, or sometimes five, forward gears. The output shaft, we can now assume, is rotating at an appropriate speed. The power is carried rearwards by a propeller shaft.

In order to accommodate deflection and movement in the body of the car along its length, the propeller shaft is equipped with flexible (or universal) joints at each end. Sometimes this propeller shaft is exposed beneath the car. At the rear axle, the driving force is split in two and turned through 90° to the two wheels.

At the end of the propeller shaft a bevelled pinion interlocks with a large crown wheel. This forms a final gearing down. By a further arrangement of bevel pinions in a box known as the 'differential', the drive shaft to the wheels are driven at the appropriately different speeds when cornering (Figure 22d). You can think of the gearbox assembly, the axle units of two half-shafts and the differential as building blocks, like the engine. Modern car design puts these large and complex standard assemblies together in slightly different ways to give different cars. For instance, all the Fords – the Escort, Capri, Cortina and Granada – had gearboxes of one type. Such standardization is very common and gives savings in manufacturing cost.

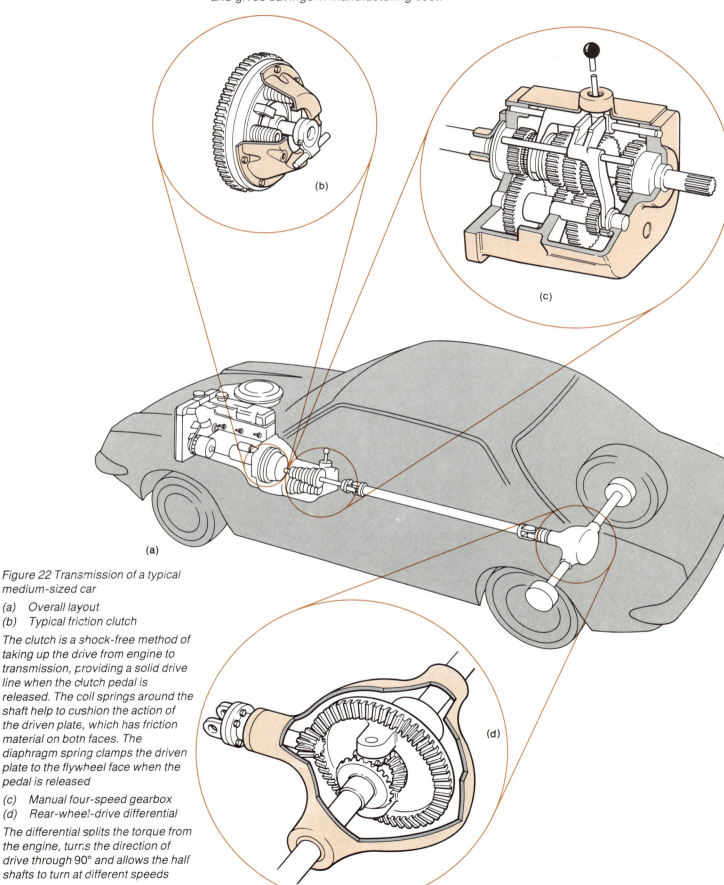

Figure 22 Transmission of a typical medium-sized car

(a)   Overall layout
(b)   Typical friction clutch

The clutch is a shock-free method of taking up the drive from engine to transmission, providing a solid drive line when the clutch pedal is released. The coil springs around the shaft help to cushion the action of the driven plate, which has friction material on both faces. The diaphragm spring clamps the driven plate to the flywheel face when the pedal is released

(c)   Manual four-speed gearbox
(d)   Rear-wheel-drive differential

The differential splits the torque from the engine, turns the direction of drive through 90° and allows the half shafts to turn at different speeds

2.2 Body shell

The development of the car in the 1920s and 1930s shows a movement away from the structural frame (or chassis) plus coachwork (or body shell) to a form of construction in which *structure and shell* were combined. The whole body of a modern car can be thought of as a stiff hollow tube braced by intermediate cross panels. 'Unitary' body construction does not mean the body shell is made of one piece, but rather that the body is conceived as one structure.

Early assembly methods seem to have been the main factor in prolonging the manufacture of chassis-built cars until the 1950s. Sheet steel had been used on European cars since around 1900. Similar alloy steels were used on the 1908 Model T Ford, the Tin Lizzy. In England the 1912 BSA was the first car to have an all-steel body. However, in America the Dodge brothers could claim to be the main promoters of pressed steel bodies from 1914. From then onwards, press tools were available in suitable sizes (Figure 23). The means were there, at that date, for the unitary (sometimes called monocoque) self-supporting, chassis-less construction, but it did not come for many years. This was partly because of the cost of huge presses, partly because other methods seemed adequate, and partly because accurate welding and framing techniques needed development.

As cars became faster and the engines heavier and more powerful, the stiffness of the simple H-chassis had to be reinforced. The sections became deep closed box sections with strong cross-members. Later still, central X-shaped bracing was incorporated, though on some cars the side members were then reduced or eliminated. The Lancia Lambda (Figure 24) went a step further with deep side members forming body stiffeners.

By the 1930s several features of cars had changed to bring unitary construction nearer. The passenger cab or body, previously constructed of wood by coachbuilders, now gave way to bodies fabricated by welding together several pressed steel panels.

The first unitary construction appeared in 1935 on the Opel Olympia and in 1937 on a British Vauxhall. However, these were little more than pressed-steel bodies welded to conventional box chassis or subframes. This demonstrated that a body properly secured to a chassis made a stiffer frame, and lead naturally to a unitary, stiff structure that provided all the requirements of passenger space, and of a mounting base for all other components. Thereafter, body and chassis were increasingly considered as a single structure.

The Second World War left the European private car industry very much diminished. First attempts at resuscitation produced car designs carried over from the 1930s. To invest in new plant needed the prospect of long runs of production, growing outputs and fewer changes. This depended on general economic recovery, which took time. By 1950 several European cars went into production with unitary body construction. The SAAB's body was formed from twenty individual pressings welded together (albeit fitted to a wooden frame).

Remember the description of the original craft-based assembly of cars before the Model T. Compare it now with modern car production. It would take me a whole unit to describe the latter fully. However, you can get some idea of the scale of the revolution if you think of the operations broken down into primitive actions. Current models would entail some ten hours of cutting, stamping, pressing and welding of sheet steel, six hours of protection and painting, and

32

Figure 23 Giant press at the Vauxhall factory around 1930

A sheet of flat metal is fed into the machine; this sheet is formed under a force of something like 400 tons, the result being seen in the foreground. Changing the dies used in presses like this make it so expensive to re-tool for a new model. Morris, for example, spent £120 000 in the late 1920s for the design and dies for his first all-steel saloon car

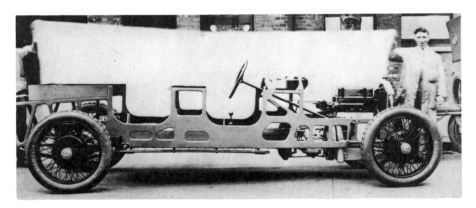

Figure 24 Italian Lancia Lambda

The chassis side-members were integrated into the body structure, anticipating modern designs. The holes are to reduce weight

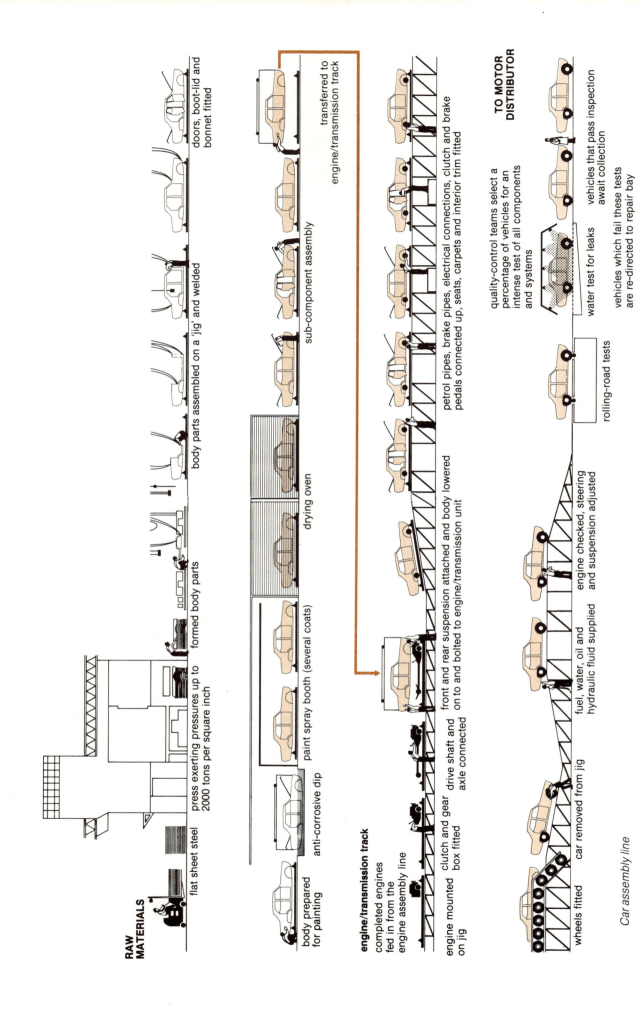

Car assembly line

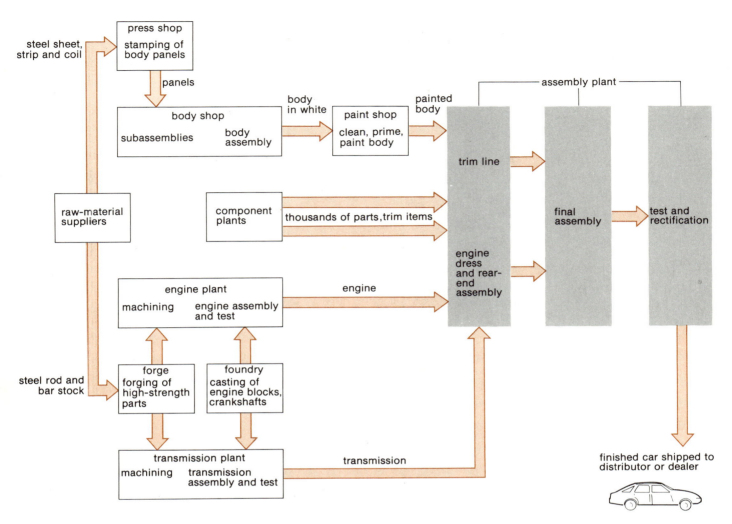

Figure 26 Sequence of operations in whole car production

six hours of various assembly operations. Of course, there are many bought-in parts and subassemblies from specialist manufacturers (lamps, exhausts, instruments, brakes, wheel drums, windscreens, door latches and hinges, upholstery, etc.). The significance of this is that 45 per cent of internal manufacturing operations involve forming flat sheet steel into three-dimensional shapes. Figure 25 shows how a modern car is made and Figure 26 is a schematic sequence of operations.

It might seem odd to you that steel sheets are still the widely preferred material for making car bodies, particularly when you think of both the resistance to corrosion and the weight advantage that aluminium offers. Occasionally aluminium has been used, in, for instance, the gleaming Gladiator of 1914 in Figure 27, but there are intrinsic difficulties. Table 2 compares the properties of steel and aluminium, steel given a base 1.

Table 2 Comparison of properties of steel and aluminium

Property	Steel	Aluminium
tensile strength	1	$\frac{1}{3}$
ductility	1	$\frac{1}{2}$
density	1	$\frac{1}{3}$
elasticity	1	$\frac{1}{3}$
cost per unit weight	1	3

Source: Fairbrother (1981) P. 18.

Notes. Tensile strength means the stress under tension at which a material fails.
Ductility means the ability of a material to undergo large deformations without fracture.
Elasticity means the ability of the material to regain its shape after deformation.

Figure 27 Burnished aluminium Gladiator of 1914 (with Jules Verne styling), breaking away from rectangular coachwork. An Edwardian view of the future car

So you can see that the disadvantage steel has in weight is offset by increased strength, elasticity, ductility and, above all, cost. Not only that, rust-proofing methods are gradually improving.

Figure 28 shows the typical range of pressings that go into a modern car (around 1981). Mass-production of these elements usually entails four subassemblies – floor, two sides and roof – which come together on the assembly line to be spot-welded to form a complete body shell.

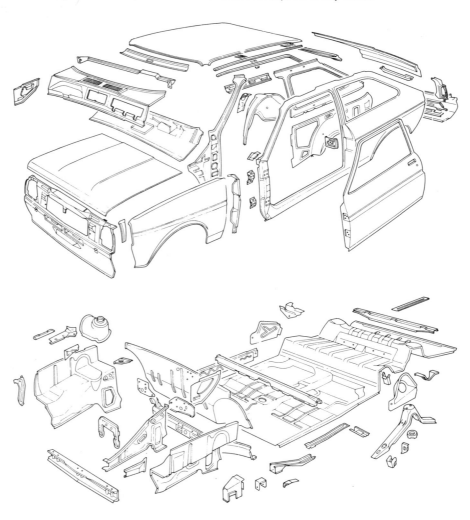

Figure 28 Typical body assembly, with seventy separate component pressings

In concentrating on the process of manufacture it is easy to lose sight of other, equally important, essentials. In general, the function of the car body is to protect the human occupants and to protect the mechanisms that propel the car. This means at one level giving structural rigidity for normal, and abnormal, circumstances. At another level it means protecting occupants from the weather, from wind and engine noise. At another level still (to do with use, rather than engineering) the body shell makes cleaning and surface maintenance relatively convenient. Imagine, if you will, the prospect of cleaning a nineteenth-century car returning from a muddy drive.

What are the main technical objectives of a car body shell?

The designers of modern body shells must attempt to meet the following technical targets. They must try:

to produce a safe compartment for the passengers;

to provide location and anchorage for all the various components;

to embody the characteristics required for effective driving – vision, sight lines, for instance;

to achieve noise and weather protection;

to permit maintenance and cleaning;

to convey a form and appearance that may actually sell the product.

Many of these factors will compete with one another. Let's look at some of them in more detail.

Safety

Providing rigidity and strength where it is needed is a fusion of experience, calculation and art. The designer has to determine: where added strength might be essential and to achieve it with bending or folding into complex sections; where to reduce weight punching holes; where to reinforce or stiffen with gussets, fillets or webs; where to corrugate to reduce flexure or eliminate vibration modes. There are computer models that help the designer, by analysing the geometries and stresses in panels after pressing. It is sometimes difficult to determine the local stretch in panels with bends in two or more planes without these computer aids. Thus it is observable that present body panels are becoming more sophisticated than, say, those of the 1950s, when press-tool designers did the job without such aids.

Also issues of safety are prescribed by national regulations, standards, and the pressure from insurance companies. For example, there are United Kingdom regulations about the amount of collapse or crumpling of the car under a frontal impact at certain speeds. These regulations represent the commoner type of medium-speed accident. The driver and front seat passenger should expect the engine not to be pushed into their laps. These conditions are met by considering the failure modes of the engine cavity structure and arranging stiffening panels, so that any collapse takes place in a predetermined direction with a calculated amount of energy absorption. This is more a case of designing failure modes and creating *controlled collapse* than of strengthening everything. The car body must strike a balance between overall rigidity and controlled failure.

There are mandatory requirements for continuous panels between the fuel tanks and the passenger compartment and a firewall or bulkhead forward of the passenger compartment. In the 1960s this bulkhead shape could trap one's knees in a frontal collision and in the absence of seat belts, could cause reverse bending of the knees and dislocated hips (apart from propelling the subject through the windscreen). In the United Kingdom the Rover car of 1963 brought in several innovations, including a steel bulkhead that deflected the engine downwards and without recesses to trap the occupants' knees in a collision.

Anchorages

There are many functional units that have to be located and anchored to the body: engine, suspension, transmission covers or torque tubes, doors, facia panels with instruments, lights and so on. Each component may demand a bracket, a mounting lug, a set of holes, a recess or slot, a clip, a rivet or a threaded insert. All these have to be incorporated in the panels that make up the body shell before it is welded up and passed down the line for assembly. The economy and efficiency of assembly will be influenced by the fixings.

Effective driving

Perhaps the most important contributions to safe effective driving are a stable driving seat and a clear view of the road to the front and rear. Also, for low-speed manoeuvring (e.g. for parking) the driver wants an even more angled view and a notion of the position of the car from the parts that can be seen.

The external shape, window angles and narrow window pillars are the main determinants of visibility. This is reflected in the press work for stiff and strong, yet narrow, window pillars. They might have to resist the weight of the car in a roll-over, so they are complex folded sections welded together. The dimensions required for stiffness are in opposition to the dimensions required for visibility (the pillars should be invisible).

Noise

Body design also largely determines aerodynamic and noise-generation characteristics of the car. The possibilities of aerodynamic shaping will be dealt with in section 6. Let's now look at noise generation.

The body noise arises from individual panels flexing in complex modes when they are excited by transmitted tyre motions, engine vibration or wind. You may have experienced the thunder generator used in theatricals; it is simply a sheet of aluminium or tinned mild steel. The natural oscillation frequencies are easily excited by shaking it. The Australian wobble board is a similar sound generator. The more plane it is, the more modes of vibration are possible. If you could fold a metal sheet in two then flatten it, and you would find that the single line of stiffness produced along the fold will eliminate much of the lower frequencies and be a poor generator even of the higher ones. This is the basis of noise elimination in body panels. Almost any pressing of panels to form a crease, curve or corrugation will prevent several of the oscillation modes. When visible creases are undesirable a strip welded to the back to form a T will create a biased stiffness that will have the same effect. Door panels may be formed with these internal bracings. Many of the stylish creases along the doors and bonnet (engine cover) have the same effect. Such creases may not conform to the best aerodynamic shape, or even what is thought to be aesthetically attractive.

Aesthetics

The aesthetics (literally, 'theory of beauty') are hardest of all to define. The overall form of the car body is very important in what it conveys to prospective buyers instantly. The form and the details of styling can convey an impression of speed, power, unity and efficiency. These are the *outward* signs of what the designers hope are *internal* characteristics. This ambiguous territory can be thought of at three levels:

the things that appeal to a particular car buyer individually;

the things that market research demonstrates appeal to large groups of people;

the things that are fashionable for a given society at a given time, that is, a very large number of people who share general views.

The first is the most difficult to deal with, in that the individual's taste might be simply odd. However, this does not matter much to a car manufacturer unless it is revealed that this individual is the tip of a, hitherto unknown, iceberg of car buyers. The third characteristic we can all observe historically. In the early 1980s we are emerging from a period of body shapes that have been angular and sharp-edged into a period of rounder bulbous shapes. Later I shall discuss the rational reasons for that particular shift. However, there may be intangible, irrational reasons beneath the development in car shapes. People, designers and consumers alike, do not feel content with things remaining the same. It is somehow offensive, as if no progress has been made. Every year cars change shape slightly, every ten years the consensus about the right shape changes, and every twenty years or so radical changes in engineering tend to take place.

SAQ 8

Can you make a note of pairs of factors that compete one with the other in unitary body design?

2.3 Audiovision Package 7: Car designers talking

At this point it would be appropriate to study the audiovision package entitled 'Car designers talking', which consists of a tape of twenty-five minutes duration plus explanatory notes and illustrations in the Supplementary Material. On the tape you will hear two designers talking about novel forms of car design.

One designer, William Towns, works as a private consultant, but his early experience was at Rootes and with Aston Martin. His work entails not just high-price saloons, such as the Aston Martin, but experimental urban cars, such as the Microdot.

The second designer, Roy Brocklehurst, works at British Leyland Advanced Technology. His recent work has been on the Energy Conserving Vehicle Mk3 (ECV3 for short).

The orientation of the two designers is slightly different; one speaks as a private specialist in styling, the other speaks as an engineer researching new directions for mass-production. Both are enthusiasts for cars and both think seriously about future problems. When listening to the tape make a note of ideas that may provide a stimulus for your own assignment on the future car.

Before you listen to the tape it might be useful to look through the Supplementary Material for your assignment, TMA T263 05, if you have not done so already. Briefly scan the Audiovision Notes also before you sit down to listen. When listening to the tape keep your Workbook to hand to make any jottings as they occur to you.

3
HUMAN FACTORS

3.1 The human element

For the moment let's put on one side the technical problems, the form of engine, the arrangement of the transmission and the construction of the body shell. These factors are the province of qualified engineers. Instead, let's concentrate on an area that is more familiar to non-specialists: the human element.

I want you to think a little about how the motor car performs, *not* as a piece of engineering, but as an artefact subject to *human use*. In Block 1 *Everyday Objects* you had some experience of evaluating artefacts on the basis of human factors. There you thought a little about why some things don't work and how they could be improved. For instance, you looked at and evaluated dashboard displays. So bring the same attitude to bear here, and apply the same principles to the whole car.

Think about the car in terms of your *own* use. You might like to make a user trip as explained in Blocks 1 and 2 in order to refresh your memory and sharpen up your observations. It does not matter if you are not a driver. In fact a passenger might have a more detached, uninvolved and unprejudiced view of the motor car. If you want to remind yourself of these techniques, go back and look at pp. 23–5 of Block 1 *Everyday Objects* and pp. 101–4 of Block 2 *Bicycles*.

From your previous experience, and perhaps using notes made from a recent trip in a car, make a list of what you think is unsatisfactory about the design of motor cars. You might find it helpful to think about this first of all from a personal viewpoint: what problems have you had with cars? Then go on to put your critical thoughts in general terms.

Here is my list. I have split it into four categories. The first three are based on my own experiences; they embody difficulties I have had in the past. The fourth category is more general.

Technical

Rust and scratch easily.
Difficult to do even simple maintenance and repairs.
Not enough internal storage space for maps, cloths, etc.
Need too many specialized tools for maintenance and repairs.
Mechanically unreliable and repairs expensive.
Ineffective windscreen wipers.
Parts from different cars usually incompatible.

Aesthetic

Elaborate but facile styling, such as stick-on trim, gratuitous insignia, unnecessarily elaborate wheel trims, door handles, etc.
Peculiar shifts in overall form from period to period, for example, sharp edge to bulbous, round back to hatchback.

Ergonomic

Variations in major controls between different vehicles, for instance, position and movement of turning indicator arm.

Driver's seat, steering wheel and pedals not sufficiently adjustable to the optimum for every driver.
Inadequate heating and ventilation.
Insufficient space to allow people to get in and out easily, or to sit comfortably inside.
Getting in and out on off-side dangerous.
Luggage handling very difficult.
Difficult to control or entertain young children.
Uncomfortable for long journeys and especially for passengers.

Environmental

Polluting.
Dangerous to other road users, to passengers and to oneself.
Cause damage to roadside buildings by vibration.
Result in environmental despoilation: underpasses, flyovers, multi-lane clearways through towns, car parks, rural motorways, etc.
Have detrimental effect on public transport.

The environmental aspects of car use that I have listed are less concerned with individual car designs than with cars as a form of transport. In fact, most of them are hidden effects and implications, which only emerged when there were sufficient numbers of cars on the roads for them to attract attention.

Pioneers such as Benz and others can have had no inkling of the enormous impact the car would have on both the geographical and social structures of society. I shall return to these more general properties of cars later. Similarly, I want to leave the technical and aesthetic aspects of design because these are dealt with elsewhere. Here I wish to concentrate on the human problems concerned with the relation between the artefact and its user, in other words the *ergonomic* aspects of design. Perhaps similar problems have been prompted by your user trip.

3.2 Space packaging

In order to pursue ergonomics in car design I want you to undertake an exercise that investigates some elements of the space 'packaging'. (This term was explained in Unit 1 *An Introduction to Design*, p. 65. It does not mean quite the same as packaging of, say, Christmas presents.)

Activity 1 Interior ergonomics

Using your Workbook and your manikins make some sketches of the way you think human beings *should* fit into cars. Give a free rein to your ideas. Take about 30 minutes and try more than one idea.

At this stage *any* arrangement *at all* is permitted. Do not worry, particularly about the size or position of the engine. You can assume it is fairly typical in size and you can vary the position.

If you want some basic elements from which to work, use the seats and internal layouts shown in Figure 29. Trace over them to give yourself new arrangements. Don't be inhibited about changing them if you think it would be beneficial to the car users.

In the list of criticisms you have already produced, one or two criticisms should act as a starting point. Focus upon one problem that irritates you. Remember, irritation is quite a *good* starting point. Designing is only a kind of scratching at an itch!

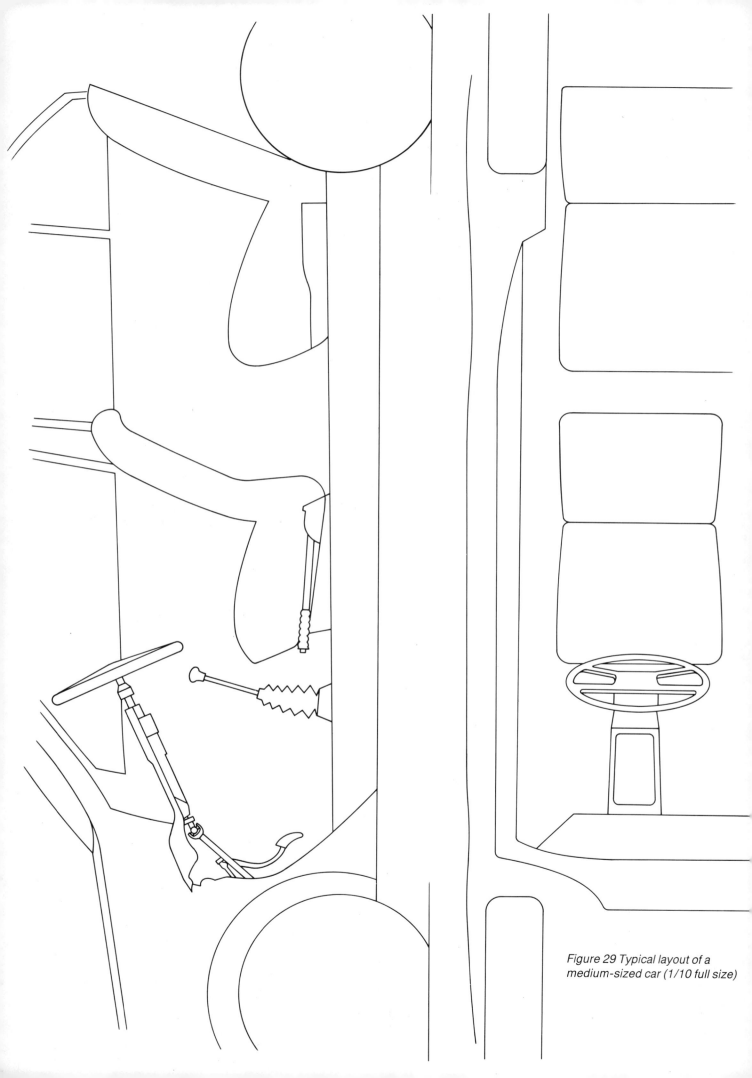

Figure 29 Typical layout of a
medium-sized car (1/10 full size)

Figure 30 shows some of the initial thoughts of the Course Team. The first problem I chose to attack concerned the relation of young children and cars.

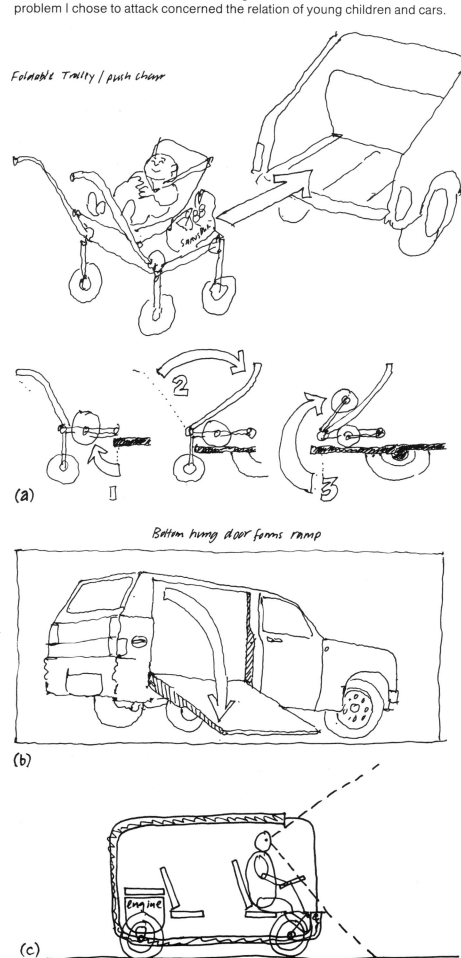

Foldable Trolly / push chair

(a)

Bottom hung door forms ramp

(b)

engine

(c)

Figure 30 Some first ideas from the Course Team

(a)   Foldable push chair
(b)   Bottom-hung door forms ramp for wheelchairs
(c)   Driver or passenger near front

At one level this means the difficulty of taking toddlers in and out of safety seats, especially if one hand is holding the push chair, the other carrying the bag of shopping, while the other is restraining another child from jumping under a truck. (All parents have three hands, at least.)

At another level it means entertaining and supervising children in the car, particularly on long journeys. My first thought was to have a rear-facing seat. This evolved to a swivelling front seat, which can face rearwards, or even sideways to allow easier movement in and out of the vehicle. I investigated this idea on plan in Figure 31.

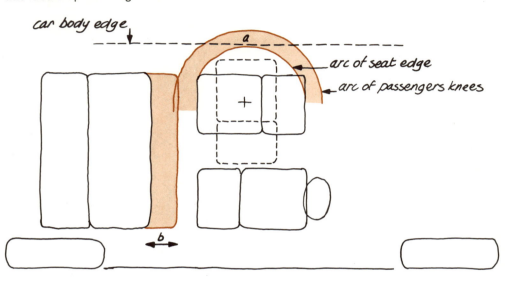

Figure 31 Plan view of swivelling front seat (1/20 full size)

Extra space is necessary at a and b

This could work, but it has two penalties. First of all, the arc inscribed by the seat front at 'a' may be made to clear the interior surface of the body, but the knees of a front seat passenger would collide with the door pillar and panel in any conventional car. Thus the width of the car at that point would have to be a few inches greater.

Similarly in the rear-facing position there is barely enough room for two sets of touching knees. Therefore the rear seat should be pushed back a distance 'b'. The increase in 'packaging' space at 'a' and 'b' is shown by the tinted areas.

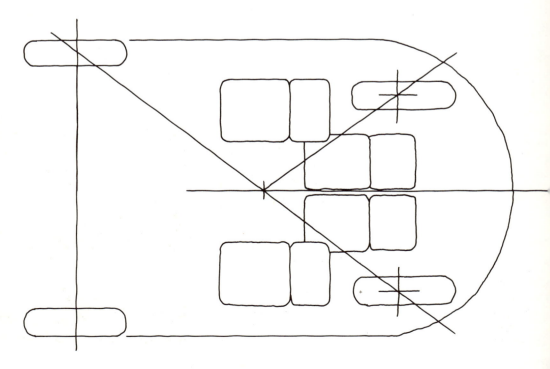

Figure 32 Widely spaced front seats (1/20 full size)

Next I tried a different arrangement of seats, which assumed an engine (or a card table) between the two front seats (Figure 32). This gave the two rear seats unobstructed views forward and made conversational exchanges easier. With this layout I tried pulling in the back wheels to reduce space. This would probably give a very odd-shaped car.

Then I looked at the arrangement in an axonometric sketch. To do this I first drew one seat carefully in axonometric projection in Figure 33. Then, having got this more or less right, I traced it for the position of the other seats in Figure 34. This shows one of the benefits of working on tracing paper.

However, at this point I remembered that I had seen such a layout somewhere before, and I tracked it down to a design by Bertone, shown in Figure 35. As you can see, I mis-remembered the original. Also, I realized that it was quite an extravagant idea, only really suitable for a mid-engined car.

Figure 33 Axcnometric drawing of a car seat (1/10 full size)

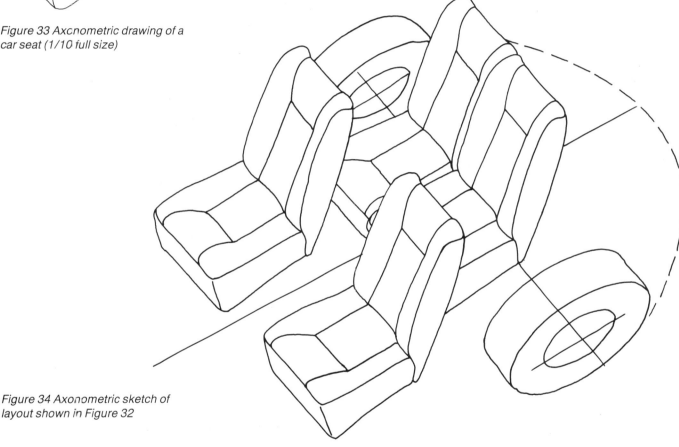

Figure 34 Axonometric sketch of layout shown in Figure 32

Figure 35 NSU Ro 80 'Trapeze', 1973, designed by Bertone

The rear seats are divided by the engine housing. The back rests of the front seats move sideways to ease the exit of rear-seat occupants

So I tried some alternative fixed seating patterns, sketched in Figure 36, and then developed the idea of a front passenger seat that can be reversed just by moving the back support forwards. In this way it could be moved to face either forwards or backwards. In either position the passenger would be able to speak to the driver, but with the seat reversed the passenger could also converse directly with other passengers in the back of the car. Another advantage of a rearward-facing seat is the extra protection it affords in the event of a crash. The design would be particularly suitable therefore for children who are too small to fit a standard safety belt.

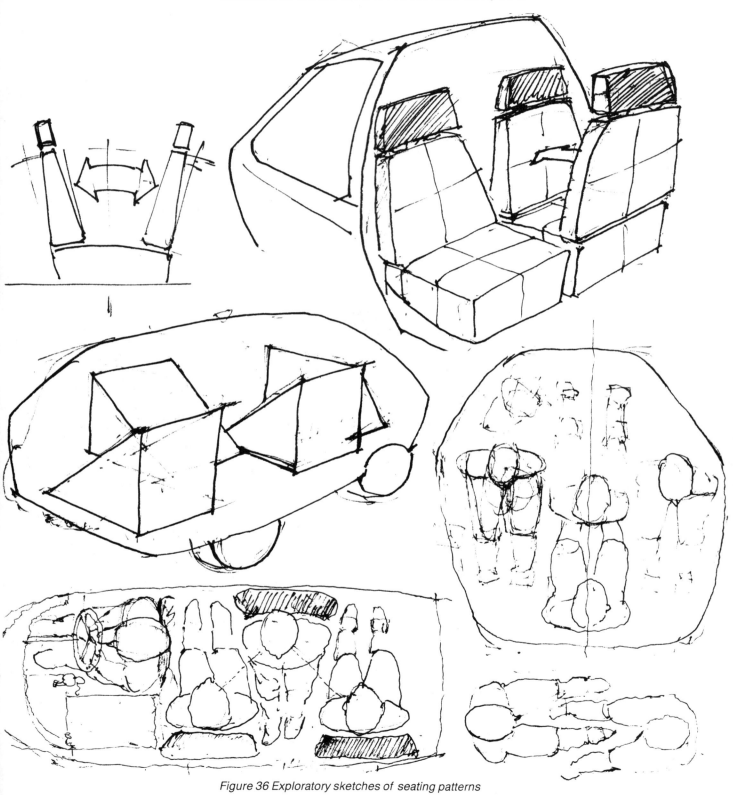

Figure 36 Exploratory sketches of seating patterns

From my rough sketches it looked as if my design might add to the overall length of the passenger space because of the extra leg room needed when the front seat is reversed (see Figure 37). Although I thought I might get away with this by making the base of the front seat slide forward when the back was flipped over, as indicated in Figure 38. However, I had now reached the stage where I wanted to work more precisely. I had the growing feeling I was allowing my sketches to fool me. For example, the driver's view might be inhibited.

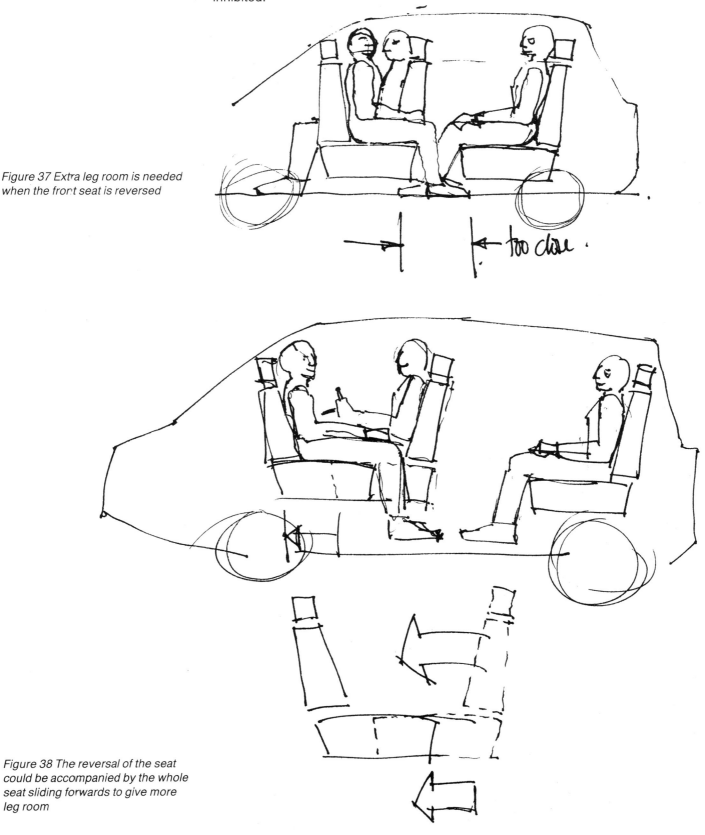

Figure 37 Extra leg room is needed when the front seat is reversed

Figure 38 The reversal of the seat could be accompanied by the whole seat sliding forwards to give more leg room

47

The *Humanscale* package in the Home Experiment Kit contains the kind of dimensions needed to check rough ideas like this. From the data on Sheet 2a *Seating Guide* I was able to construct a space envelope for the driver of an ordinary saloon car using the dimensions shown in Figure 39. A space envelope is an imaginary bubble around a person that delineates the critical dimensions of the amount of room needed for a human activity of some kind. In this case the human activity is driving a car and the dotted line in Figure 39 shows the minimum amount of space required by a driver to allow for reach to dashboard, head clearance, comfortable sitting posture, etc.

Since my main concern was to check whether my new seating arrangement added significantly to the overall length of the car I selected measurements appropriate to the 97.5th percentile male to produce a side elevation. From this I was able to trace another silhouette to represent a passenger, which I then laid over the first drawing to give the conventional seating arrangement shown in Figure 40. The important dimension to note in Figure 40 is that labelled 'H-couple'. This is the horizontal distance between the hip pivot points H of the two seated figures. H-couple is given by *Humanscale* Sheet 2a as 965–1041 mm, and I needed the top end of the range since I was designing for the largest members of the population. From Figure 40 I was able to measure off the minimum overall length of the car interior as 2530 mm.

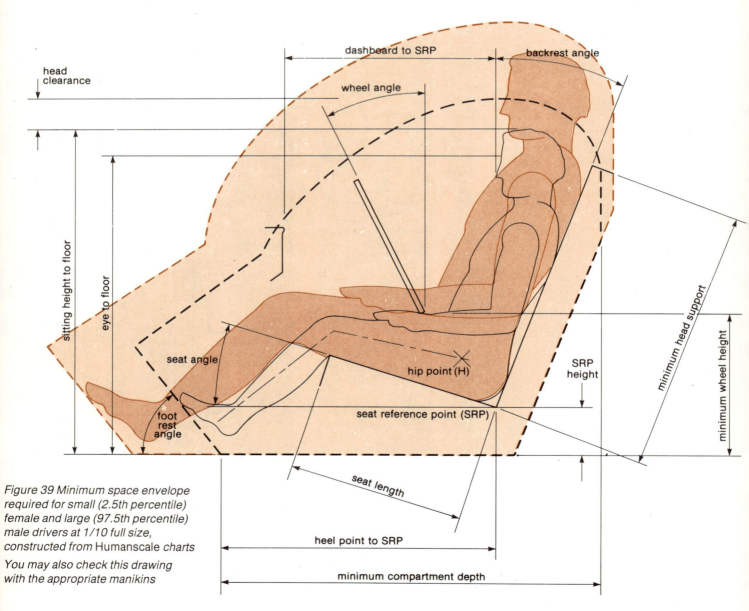

Figure 39 Minimum space envelope required for small (2.5th percentile) female and large (97.5th percentile) male drivers at 1/10 full size, constructed from Humanscale *charts*

You may also check this drawing with the appropriate manikins

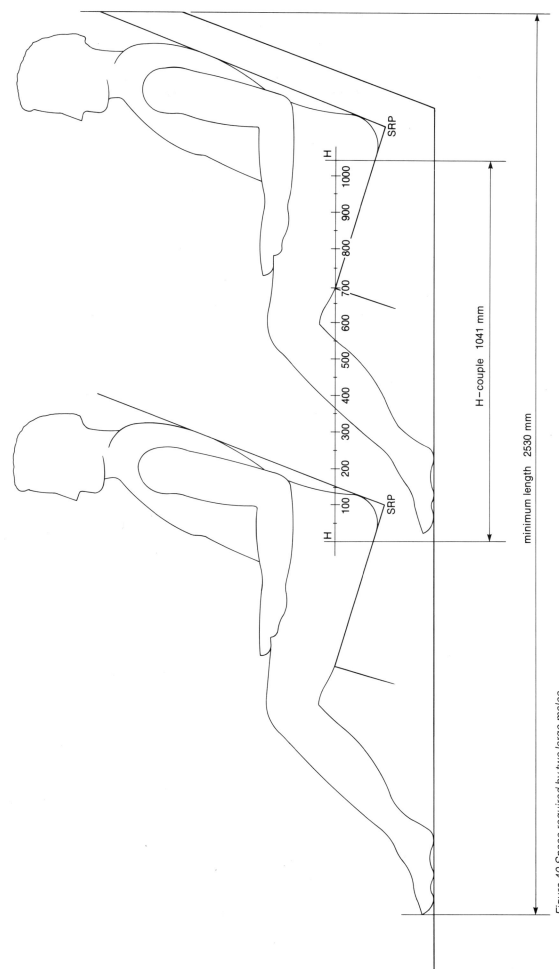

Figure 40 Space required by two large males

The next step was to compare my reversible seat design with this minimum standard. By making another tracing of the passenger and reversing it over Figure 40 I was able to see exactly how well a reversible seat would fit into the available space when occupied by a large passenger. It did not fit at all well. To provide enough leg room for the rearward-facing passenger I should have to move the back seat back by at least 250 mm and even then passengers facing each other would have to sit with their legs interleaved.

To provide as much leg room for all passengers as they could expect from a standard seating arrangement I would have to add 1120 mm to the length of the compartment which is almost as much as it would take to add another row of conventional passenger seats. I was surprised to find that the arrangement added so much to the overall length of the compartment. So I decided to make some comparisons with existing cars. Table 3 gives measurements for the distance defined as H-couple for several popular small cars.

Table 3

	Length of H-couple
Austin Metro	734 mm
Ford Fiesta	725 mm
Renault 5	749 mm
Fiat 127	727 mm
Volkswagen Polo	684 mm

Source: *Motor*, 18 October 1980.

As you can see, none of these cars has an H-couple even approaching the *minimum* recommended *Humanscale* dimension, let alone the maximum figure I used to allow for larger passengers. So although it seems that conventional seating *arrangements* are hard to beat in terms of their efficiency, it looks as if there is considerable scope for improvement in the amount of space allocated to car occupants.

The purpose of this activity was to emphasize that design goals are rarely simple. Smallness is not everything, but on the other hand it is an important factor in modern car design because of the materials costs and weight implications of large cars. Design solutions therefore tend to take the form of trade-offs between alternative requirements of comfort and economy, safety and technical performance and so on. At present there is a trend in favour of small cars, so the balance is tipped against factors such as comfort. Nevertheless the trend towards even smaller and more compact cars means that it is increasingly important to know as precisely as possible just how much space people need.

In 1890, when Panhard and Levassor designed their first car, they perched the driver and passenger on bench seats out in the open. They did not have to concern themselves with detailed or precise anthropometric measurements. Fitting the occupants into modern cars by comparison is a much more skilled packaging exercise, for which detailed specifications are required just as they are needed for technical and economic aspects of the design.

PART TWO

4

HOW CARS ARE DESIGNED

4.1 Introduction

The purpose of this section is to describe in some detail exactly how modern mass-produced cars are designed. More than that I shall try to show how various levels of complexity are encountered as designers converge towards a single artefact, or family of artefacts. At the beginning the design process is simple, cheap and wide-ranging. The process becomes gradually more complex, more expensive and narrow in its targets. The account of the design process given here follows this convergence chronologically.

At this point it would be helpful if you were quickly to re-read section 4 of Unit 1 *An Introduction to Design*, which gives a general outline of some of the procedures involved in car body design.

Can you summarize the main points of section 4 of Unit 1 *An Introduction to Design*?

This is my summary of the main points from Unit 1:

A new car takes between three and six years to initiate, design and develop.

These three stages involve anything from a hundred to over five hundred professional engineers, stylists and project planners, plus a large number of specialist craftsmen.

Such a project requires the investment of a huge amount of money, generally tens or hundreds of million of pounds.

A lengthy preliminary investigation is conducted and an elaborate decision-making process determines the exact size and type of car to be produced.

From these deliberations an agreed 'package' is derived, and a car body designed around it. The basic shape of this body is the result of a further lengthy process of evolution and refinement involving clay models, or 'styling bucks', and wind-tunnel tests.

Measurements taken from the finalized styling buck are used to construct the first steel prototype.

In designing the car and its structure, a range of computer-aided techniques are now widely used for a number of purposes, mainly to achieve the greatest strength for the least weight.

Prototypes and 'pre-production' cars are subject to a long programme of testing and development, both on the road and in the laboratory.

In Unit 1 also I talked of 'design convergence'. You may have felt at that time that this was just a loose general expression to convey the idea that designers start from a lot of ideas and end up with *one* product. However, I can be a little more precise than that. This section describes the techniques used to effect that design convergence. But firstly, what does the overall process look like when applied to the car industry?

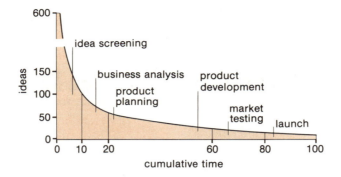

Figure 41 Mortality of new product ideas

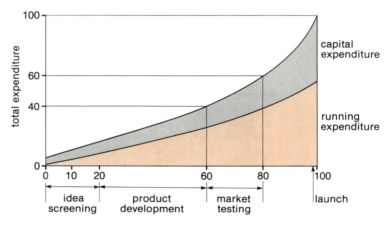

Figure 42 Cumulative expenditure during the design process

It is possible to make an idealized plot of the number of ideas against the design programme in time (Figure 41). From the first screening of ideas, through market analysis and product planning, two or three options, or more accurately a single family of related options, survive. New ideas at the beginning are cheap, but even trivial changes to the final product are expensive. This can be represented by Figure 42, a similar plot of expenditure against time, which is the mirror image of Figure 41. At the end of the process all the money for development has been spent, for instance, £700 million for the Ford Sierra. At the beginning of the process expenditure is low: only on existing capital equipment and the time to turn over ideas; one or two people and a few pieces of paper. Let's examine the first beginnings.

4.2 What kind of car?

Section 4 of Unit 1 *An Introduction to Design* covers, to a limited extent, the way in which a company's managers decide what size and kind of car should be introduced. This process is so intimately bound up with the actual design of the proposed car that it is necessary to enlarge upon it here. In fact the planning of a new product seldom takes as long as it did for the BL Metro referred to in Unit 1: that project was dogged by politics and sudden reversals of fortune to quite an exceptional extent.

Thus, unlike BL during the early development of the Metro, most of the motor manufacturers have a stable management regime. For such companies the first major factor influencing the decision to launch a new car is very often nothing more than the conviction, or 'gut feeling', of the directors, perhaps prompted by declining sales of a particular model. In the late 1950s, for example, the managers at Rover came to the conclusion that the long-term survival of their company depended upon a move down-market from luxury cars like the $3\frac{1}{2}$ litre model and the rather staid Rovers to smaller, cheaper and more widely accessible models. The result was the Rover 2000 introduced in 1963. Much more recently, the management of Daimler–Benz in West Germany arrived at a similar conclusion and thus introduced the new 'small' Mercedes (Figure 43). Such designs are premeditated moves into a new market sector.

Figure 43 Small Mercedes, introduced in 1983

The second important factor is feedback from the sales organization and the dealers. 'If only we had a small model to start people off with,' they'll say, or 'If only we had a large model for customers to "trade up" to.' The salesmen may perhaps see a particularly successful model made by a rival manufacturer and ask that their own company's range be expanded to challenge a competitor. Thus that highly successful sporting hatchback, the Volkswagen Golf Gti has spawned a number of rivals, among them the Ford Escort XR3.

Nowadays the big motor manufacturers also pay a good deal of attention to a third major influence: market research. Many cynics believe that market research is only as good as the person interpreting it. It is certainly true that if the senior managers lack a good understanding of the motor car – which, after all, is a pretty subtle and complex machine, part consumer durable, part cult object – then the results can be disastrous.

Therefore a less well-defined and less controllable factor is the role of the *product champion*, the individual from whom, or around whom, the main ideas coalesce. This was clearer in the past: Henry Ford and his Model T, Herbert Austin and his Austin Seven. But similar cases are Issigonis and the Mini, perhaps even Michael Edwardes and the Metro. Bob Lutz, the outgoing Chairman of Ford Europe, has made an interesting comparison between the car industry and the film industry.

> There's a very fine line between doing a movie that gets out there and fails, and doing a movie that's *Star Wars* – and yet the celluloid's the same, the actors are the same and so is the amount spent on special effects. One film's good, the other isn't, because there is that creative and psychological content in any product programme that defies a totally systematic approach.
>
> (*Car*, October 1982).

The comparison is appropriate because both industries entail a huge investment of money, elaborate technology, creative individuals, teamwork and an underlying uncertainty about what the public will accept: a kind of high-level gambling. In order to sustain a series of ideas through to successful completion, a particular kind of determination is needed. This may come from an executive, a designer or an engineer. Thus we can identify the Ford Cortina with Terence Beckett, the Ford Fiesta with Iacocca, and the Sierra with Bob Lutz. This is a matter of one individual's stamina as well as market assessment.

Nevertheless, intelligently conceived and cautiously interpreted market research undoubtedly has considerable power to identify unsuspected groups of potential buyers and gaps in the market. It was market research, for example, that identified a large socio-economic group in America, mostly composed of young but prosperous married couples, often with small children, and all wanting a smart sporting car rather than a staid saloon. Ford introduced the Mustang to satisfy the motoring needs of this group, and very successful it has been, while the European equivalent, the Capri, hasn't done badly either.

At the same time, there's a good deal of substance to the suspicions of the sceptics. To begin with, the basic assumption of market research is that people actually know what they want, when often they don't, until they see it. Moreover, market research tends to look at *past* trends and to assume that they can be reliably extrapolated into the future. Unfortunately, though, there's very little correlation between the past and the future, especially in the car business. Thus market research would *not* have predicted the way in which the layout of the Mini with transverse engine and front-wheel drive revolutionized car design in twenty years, nor would it have foreseen the popularity of hatchbacks before the arrival of the Renault 16, which pioneered them.

Hence the sort of new car to be launched is essentially determined by three basic factors: the experience and beliefs of senior management; the feedback from the sales organization; and market research. To study these parameters with the care needed, all but the smallest motor manufacturers now have a product planning department and a product planning committee, which includes senior managers such as the chief engineer, chief stylist and finance director. Among the duties of the product planning department will be continuous monitoring of the motoring scene, and constant evaluation of rival cars. For instance, at the time the Mini came out, Ford UK bought one, took it apart, costed every item and decided that for them such an excursion into the extra-small car market would be unprofitable.

Once the basic germ of an idea has been formulated, the precise size and price-class of the proposed new car must be determined. This is a more complex business than it seems, and can lead to disaster if not properly considered. Remember, for example, the Vauxhall VX of some years ago (Figure 44). Despite being a handsome, comfortable car with many virtues, it never sold as well as it should, almost certainly because it was too big for the Cortina sector of the market and too small for the Granada sector. In the eyes of many prospective buyers, it was neither one thing nor the other.

Figure 44 Vauxhall VX

Ford, in fact, have five different classifications for the various sizes of cars.

The width of a car, incidentally varies relatively less than the length: the smallest cars are around 5 feet wide and the largest around 6 feet wide.

Changes in the pattern of demand over the years can outdate classifications of this sort, which is why a special 'intermediate' C/D category had to be created for cars of the Cortina/Sierra type, despite the tremendous importance to Ford of this sector of the market over the past two decades. Other manufacturers have similar classification systems, some of them elaborate.

Size, of course, is by no means the only determinant of a car's market appeal: its 'image' or character may be equally important. At various times, for example, different manufacturers have sold quality small cars of the Triumph Dolomite sort at prices far higher than those appropriate to ordinary cars of the same size, and to quite a different type of customer.

While a motor manufacturer's product planners might indeed consider launching a new car to satisfy such a particular and specialized need, they are far more likely to seek a type and size of body shell that will suit the widest possible range of engines and levels of luxury and interior trim. In fact the need to sell large numbers of a given body shell to recover the very high cost of its development and tooling makes such an approach essential.

Having decided upon the size, price-class and character of the car to be introduced, the next step is to choose the kind of body required for it. If the proposed car is to be, say of the Ford Capri kind, then the type and style of body will essentially choose itself: it will have to be a sporting 2 + 2 with some sort of fast-back shape. In other sectors of the market, however, the choice is wider. Thus the Rover 2000/3500 and Renault 20/30 have hatchback bodies, but compete against such rivals as the Peugeot 604 and Ford Granada, which have 'three-box' saloon bodies with a conventional protruding boot at the rear.

Here again the tendency is for the modern motor manufacturer to produce as many variants of a given car as possible. Thus Volkswagen produces both a three-box version, the Jetta, and a coupé version, the Sirocco, of the Golf hatchback (Figure 45), while Talbot makes the Solara, a 'booted' version of the Alpine hatchback.

Figure 45 Variations on a theme

Volkswagen's hatchback Golf, notchback Jetta and coupé Sirocco are all built around the same basic floor pan and mechanical units

When the product planners have decided on the type of body to be used, they will have built up a very clear picture of the proposed new car and will have confidence in its future commercial success. Now they will have to decide whether their company can afford to build such a car or not. Practically every motor manufacturer in the world, however large, is constrained, when considering a new model, by prior, often heavy, investment in existing engines, transmissions and suspension systems. *All-new* cars like the Wankel-engined NSU Ro80 and the Alfa Romeo Alfasud are very rare indeed.

Existing commitments, therefore, can have a profound influence on the design of new models. Thus, when almost every other manufacturer in the world was turning to front-wheel drive for models of every size, Ford returned to rear-wheel drive for the Sierra, largely to accommodate a wide range of existing in-line four-cylinder and V6 engines. Even when adopting a completely new design approach, most companies tend to introduce a new car that is only *partly* new, having, say, a completely new body shell and suspension system, but a 'carryover' engine and gearbox.

What is possible, therefore, in the light of existing commitments, may be rather different from what is theoretically desirable. When the two have been reconciled, the product planning department will be in a position to produce a dimensioned 'package' drawing (Figure 46), which specifies in detail a wide range of parameters such as leg room, head room, boot space, engine compartment size, suspension location and so on.

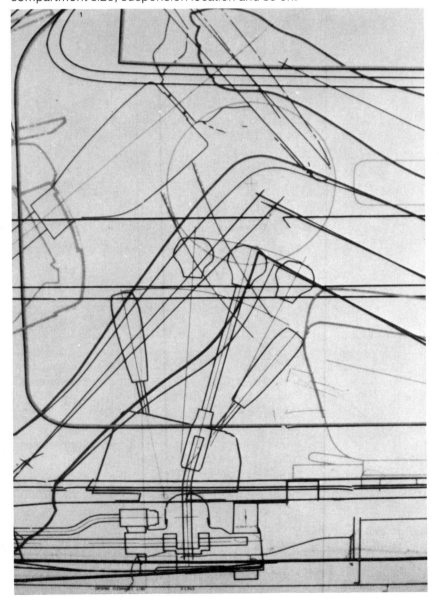

Figure 46 A package drawing provides the fundamental dimensions around which the stylists, aerodynamicists and engineers evolve the body shape

Detail of the package drawing for the BL Metro showing position of mechanical components and human occupants

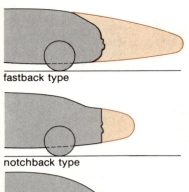

fastback type

notchback type

squareback (Kamm) type

Figure 47 Fastback, notchback and squareback

Common forms of recent motor cars

At roughly this stage in the project the company's planning staff will lay out a timetable for each stage in the design and development of the new car. (See Figure 39 on p. 66 of Unit 1 *An Introduction to Design.*) So complex is the design and development process, that rigid adherence to the timetable is essential if the car is to be introduced on time. Hence a delay of even a week at any stage would be regarded as quite unacceptable by senior management, even if the programme still had three years to run, and even if the part involved were relatively minor.

The chief stylist will have given advice at each of the previous stages, but up to the mid-1970s, it would not have been until after the general specification and package had been agreed that his major task began: to design a pretty suit of clothes, as it were, around the specified dimensions.

Nowadays, however, preliminary wind-tunnel tests of simple scale models are often conducted at an early stage, either to discover the best basic shape for the likely package or to make sure that a shape and style of body already preferred is capable of development to a low enough level of aerodynamic drag. Almost from the start of the Sierra project, for example, Ford aerodynamicists measured the drag coefficients of a wide range of simple $\frac{3}{8}$ scale models, which were provided with detachable tails and noses to facilitate testing. Fortunately for everyone concerned, the 'aeroback' shape — half fastback, half notchback — which the stylists already preferred was found to have the necessary aerodynamic potential. The Ford engineers insist, though, that, had they found a completely different but significantly better shape, they would have adopted it.

Aerodynamic drag is thus yet another factor to take into account and when the product planners have allowed for its influence, the final result is a revised and agreed package drawing specifying a car:

that is of the precise size, carrying capacity and body type required;

that uses such existing assemblies as the company's senior management considers necessary;

that is of a basic shape likely to allow a low drag coefficient for the size of the car involved.

SAQ 9

What are the three starting points for a new car described here right at the beginning of the design process?

4.3 Styling

From the package dimensions the stylists begin their attempt to create an attractive, gracefully proportioned and neatly detailed body that meets all the basic requirements. They generally start with small 'renderings' or coloured drawings (Figure 48), which are submitted to the product planning committee for approval. From these renderings the product planning committee make for further development a selection of the alternative body styles that they consider to be the most promising.

But a motor car is above all a three-dimensional object — a piece of sculpture, in fact — so the next step is to turn these renderings into a series of solid models, called 'styling bucks' (Figure 49). Traditionally, these are made of a special fine clay carried on some sort of framework or armature, often of wood. A few small-scale models may be made, for example $\frac{3}{8}$, $\frac{1}{5}$ or $\frac{1}{4}$, but full-size models are developed quickly. The first television programme, 'Reality is too much', showed some of the techniques of full-size modelling.

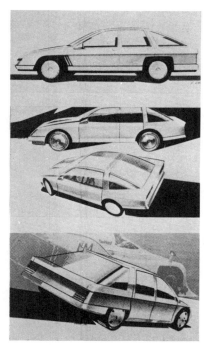

Figure 48 Car stylists begin with small coloured drawings, or 'renderings'

Figure 49 The basic shape of the car is built up with a series of models or clay styling bucks at full size

Figure 50 Tape drawings are used to refine the body shell and interior

Since the package has already been pretty carefully defined, however, a full-sized model of the proposed interior or 'seating buck' will be built to check leg room, control location, interior trim materials, etc. Refinement of the package, throughout the project is facilitated by the use of full-sized drawings, in which the lines are formed by easily-altered strips of narrow tape (Figure 50).

From the initial selection of clay models, the choice will be narrowed to a small group of these considered to be the most promising, called the 'pre-programme alternatives' in Ford's jargon. These will be further refined, both aesthetically and functionally, with the help of more wind-tunnel tests. To ensure that the tests give accurate and meaningful results, the aerodynamic models used are far more detailed than before, and faithfully simulate every part of the body being studied, including underbody components such as the exhaust system and suspension units (Figure 51). Since fine detail cannot be reproduced in clay at the ⅜ scale commonly used, and since clay tends to flake off in fast-moving air, these models are often made of an alternative material such as 'epowood', a mixture of epoxy resin and sawdust.

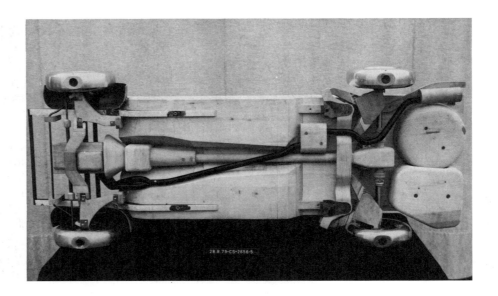

Full-sized styling bucks are then made of the one or two 'pre-programme alternatives' that both have low aerodynamic drag and are considered to be the most appealing. Eventually a final choice has to be made, and the product planning committee decides that one particular model shall be the basis for the car that is to be introduced. In Ford's terminology this is the 'go with one' decision, and when it has been made the model form will be frozen, only minor alterations then being permitted.

However, considerable further development is not only permitted but regarded as vital, especially development of the new shape's aerodynamics, since scale-model wind-tunnel tests can be very misleading. So an essential part of the programme will be measurements of drag coefficient, lift, side-wind stability, etc. taken from full-size models, perhaps mounted on a real chassis and wheels, complete in every detail (Figure 52).

By this time a good deal of development work will have been done on the various mechanical assemblies to be used. To prove their reliability and mutual compatibility they will have been fitted to modified existing cars called 'mules' so that engine, transmission and running gear can be subject to many thousands of miles of usage. The space and mounting points they need, and their cooling requirements, will all have been assessed and allowed for in the shape of the model nearing completion.

At last, though, comes the time when the functional efficiency and aesthetic merit of every conceivable detail have been agreed and incorporated into a finalized styling buck. From numerous measurements of this styling buck, the first steel prototype is built.

To take these measurements, Ford uses a 'scan mill' a large gantry-like affair (Figure 53) which can be passed over the model from one end to the other. When performing its 'scan' function, its sensitive probes register the locations of hundreds of points on the model's surface and convert them into the coordinates of a three-axis frame of reference. To eliminate any inaccuracies or imperfections built into the styling buck by the modellers, these electronically stored coordinates are subjected to a computerized smoothing process. Using these corrected coordinates, the device is passed over the model again, but this time as a mill, which shaves protuberances off the clay surface so that all the curves are smooth and accurate.

To ensure that no subtlety of proportion has been lost by this smoothing process, a glass fibre mould is made from this corrected clay model, from which in turn another clay model is made. Only when this further clay model meets the approval of all parties concerned, are its measurements taken as being satisfactory for the construction of the first steel prototype.

Figure 52 Final wind-tunnel testing involves full-size models

The man here is holding a 'smoke wand', which makes the air flow visible

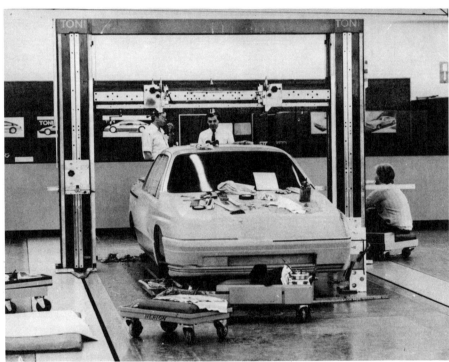

Figure 53 Ford's scan mill both measures and corrects the final styling buck

4.4 Body design

As we have seen in section 2, forty and more years ago the typical motor car was held together by a chassis frame (see Figure 15) composed of open-channel girders and quite separate from the body, which contributed little to the overall strength. These chassis were fairly good at resisting one of the most important loads to which a car is subject: the bending load imposed by the weight of the body, engine, gearbox, passengers, luggage, etc. But they had a very poor ability to resist another very important kind of load: the load in torsion created when one corner of the structure is deflected, as it will be, for example, when one wheel passes over a bump.

Since a tube is very much stiffer than an open channel of similar dimensions, some improvement in the overall torsional rigidity of a girder chassis could be achieved by boxing in the open channels, and by other measures such as the introduction of cross-bracing. In the 1930s engineers began to realize that a far better solution would be to make the whole car a gigantic tube, the floor,

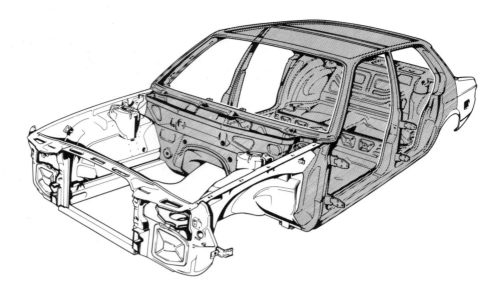

Figure 54 Modern unitary construction body shell of a BMW

Every part contributes to the overall stiffness

sides and roof of the body forming its walls. This led to the unitary, or stressed skin, form of construction, in which both body and floor pan contribute to the strength, forming an integrated whole. A modern body shell of this sort (Figure 54) has box-section side-members not unlike the longerons of the old-fashioned separate chassis, but these are only two of many structural elements, which include the floor pan, the sides of the body, the roof and the scuttle (the raised structure in front of the driver upon which the instrument panel is mounted).

The engineer designing such structures starts the job quite early in the project; it overlaps the work of the stylists. A good deal of preliminary work can be done as soon as the package drawing is agreed, remembering that this effectively defines not merely the passenger accommodation but also the nature and location of the engine, transmission system and suspension units. Between the development of the final styling bucks and the starting point in the manufacturing programme, the body engineer receives more precise information to work on when a set of 'hard points' is approved by the product planning committee. These 'hard points' are the mounting points for the engine, steering, springs and suspension linkages, and also points that define other areas critical to structural strength such as the door and tailgate apertures and the locations of the windscreen and rear window. The hard points cannot be changed once approved, so that any further modifications to the body must be confined to the outer skin or areas having little or no effect on structural strength.

Today's body designers are greatly helped by a number of computer-aided design techniques, of which the best known is *finite element analysis*. This involves resolving the sheet-metal structure into an equivalent network (Figure 55) of simple struts, or finite elements, that act only in tension or compression and hence have easily defined characteristics, mathematically speaking. A computer is then programmed with the equations of all these struts and of the connections between them. The net result is a computer-simulated theoretical structure that can be modified at will, thus minimizing the expense and trouble of much cut-and-dry experimentation.

The simulation will only be accurate, of course, if it incorporates a large number of finite elements, but the improvement here has been spectacular in recent years. Thus the structural network built up for the front-wheel-drive Escort, introduced in 1980, contained 3200 finite elements; that for the Sierra, introduced three years later, was composed of 11 574 elements. Finite element analysis leads to greater structural efficiency, that is, improved stiffness for less material. Much programming time is needed to represent a

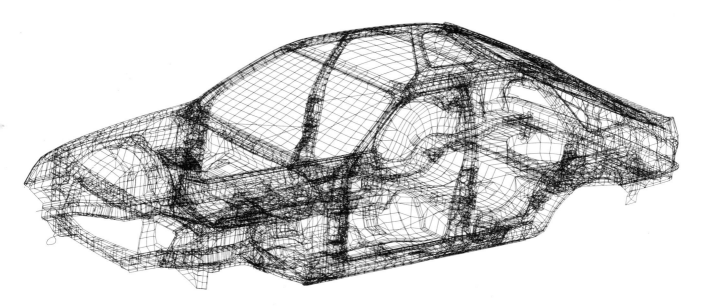

Figure 55 Structure used for finite element analysis of the Ford Sierra, showing the complexity needed for an accurate simulation

body shell in such detail. When it is completed, however, the engineer can easily modify the resultant structure – through the computer's keyboard – and swiftly learn the effect of adding a cross-member here or deleting a frame there. The construction of a real body shell then becomes a matter of confirmation rather than investigation. More importantly, the design of the strongest possible structure with the least weight is greatly facilitated.

The latest and most sophisticated finite element analysis programmes carry the design process one stage further for the body engineer. They provide dynamic as well as static information and can be used to predict the unwanted resonances, leading to unpleasant booms, that might be excited by engine, transmission or other vibrations, so that the shapes of the floor pan or body panels can be modified to eliminate them.

However, the finite element analysis technique's greatest value lies in its ability to save weight. Nowadays saving weight is just as important as minimizing aerodynamic drag and for the same reason: to reduce fuel consumption. Special lightweight, high-strength materials, therefore, such as HSLA (high-strength, low-alloy) steels, are increasingly being incorporated in the highly stressed areas of modern body shells. Ford, for example, used 80.6 lb of high-strength steels in the body shell of the Sierra. And the weight saving this achieved – 13.6 lb in the hatchback body – shows the lengths to which the modern body designer is prepared to go. In a similar bid to save weight, Citroën not only used high-strength steels in the BX model, but also used plastics (polypropylene) for the bonnet and tailgate as well as the bumpers and a number of smaller body parts (see Figure 71).

The modern body shell does suffer from one disadvantage: it is not composed of a small area of thick metal, as was the girder chassis that it replaced. Instead it incorporates large areas of thin metal and hence is very prone to corrosion. Corrosion protection in cars has made giant strides in recent years, one of the main weapons being systems by which an even layer of primer can be electrolytically deposited on the most inaccessible nooks and crannies of the body.

Some anti-corrosion measures can be taken at an early stage in the design of the body shell. Its structure must be carefully studied to ensure that water traps are not included, and enclosed box-section members must have suitable placed holes so that primer can reach their inner surfaces. Some companies, such as Volvo, provide ventilation for the sills of their bodies to discourage the accumulation of moisture within them. At a later stage wax will be injected into cavities, a PVS coating added in places, and stone chip protection provided by such means as plastic liners for the wheel arches.

4.5 Development and testing

The automotive engineers of today must do much more than merely design a rigid, rattle-free structure with a good resistance to corrosion: they must also provide its occupants with the maximum protection in an accident, and to an extent that is now minutely specified by a host of strictly enforced regulations. In essence this is done by giving the car a strong central passenger compartment, to retain the occupants and protect them against intrusion, with crushable, energy-absorbing ends, to minimize the deceleration caused by a collision. A typical modern car has to deflect by several inches, for example, during the head-on crash into a solid concrete barrier at 30 m.p.h. to which, by law, the prototype must be subjected.

Unfortunately, finite element analysis isn't of much use here, because it concerns itself with the elastic behaviour of a structure: its behaviour before the metal gives or yields under stress. A structure's resistance to crushing, however, is determined by the behaviour of sheet metal in 'plastic deformation', beyond its elastic limit. Through the careful study of barrier crash tests, analytical techniques have been developed and have been translated into computer language. As a result it has recently become possible to predict, with a fair degree of accuracy, the energy-absorbing characteristics of a car's front or rear end in a collision. This relatively new computer-aided technique is by no means perfect, but it does save a good deal of time and money. It means, say, that only six experimental prototypes have to be crash tested, compared to twenty before.

After the first steel prototype has been built, between six and twelve further prototypes will be constructed for the thorough testing of the body shell and all the mechanical assemblies. These days much of that testing can be done in the laboratory. One way, for example, is by using hydraulic suspension vibrators (Figure 56), which pound the wheels of a prototype from below as if it were running over a rough road. These can be used for fatigue tests, when they can be run all day and all night, and for NVH (noise, vibration and harshness) investigations. Controlled by computer, these vibrators can be programmed with the profile of an actual stretch of road known to excite some particular resonance.

A more important tool, indeed an essential one, is a device involving a pair of rollers that are turned by a car's driving wheels (Figure 57). This roller dynamometer, as it is called, can be used to measure the engine's power output, but more often it's used to simulate on-the-road driving conditions, which it can do with some accuracy since such factors as aerodynamic drag, the car's inertia and the rotary inertia of its transmission system can all be taken into account. Using a roller dynamometer, therefore, a car can be put through a driving cycle or prescribed sequence of idling, acceleration, cruising, deceleration, etc. It is from measurements taken during these driving cycles – the ECE Urban, the US Highway, etc. – that official fuel consumption figures and exhaust pollution levels are determined. Since cars can only be sold if they comply with the legal exhaust emission requirements now demanded by almost every country in the world, the introduction of a new model means hundreds of hours of roller-dynamometer testing. In some cases cars are put through 50 000 mile endurance tests using computer-controlled servomechanisms in place of a human driver.

A further, very useful, but rather expensive piece of development equipment is a climatic wind tunnel: one in which both the temperature and moisture content of the air draught can be very carefully controlled so that the car can be roasted or frozen and can be blasted with rain, sleet or snow. In this way the effectiveness of the heating and cooling systems in every type of climate can be fully assessed.

Laboratory testing is never enough, so a good deal of further prototype development is done on a private test track, or proving ground (Figure 58).

Figure 56 Suspension of a prototype being pounded by four hydraulic vibrators to simulate rough-road running

Figure 57 Car on a roller dynamometer

The large fan supplies cooling air

These proving grounds – all the large motor manufacturers own one – have a wide range of special features, such as cobbled, undulatory and other surfaces to assess ride comfort, banked tracks for high-speed running, water splashes, twisty circuits for the assessment of handling and road holding and so on. They are generally used in two basic ways: for investigatory tests and for long-distance endurance tests.

Nevertheless, it is relatively easy to design a car that suits a particular test track very well but isn't so happy in the real world outside. So millions of miles of testing on ordinary public roads are essential, and in extremes of climate ranging from the Arctic to the Sahara Desert. The final stages of the project, this open-road testing, will involve a hundred or more pre-production prototypes, built from the dies, tools and welding equipment that will be used when the car is produced in volume. The pre-production prototypes are often disguised with dummy structures of canvas or plastic, to foil inquisitive cameramen.

This is often a period of anxiety for the engineers, since a car built on production tools is seldom quite the same as the original hand-built prototypes. This is when doors and windows are found to fit poorly and when unwanted vibrations appear. Further last-minute modifications are often required.

And when the car does go into production, the public always seems to find faults that the engineers overlooked.

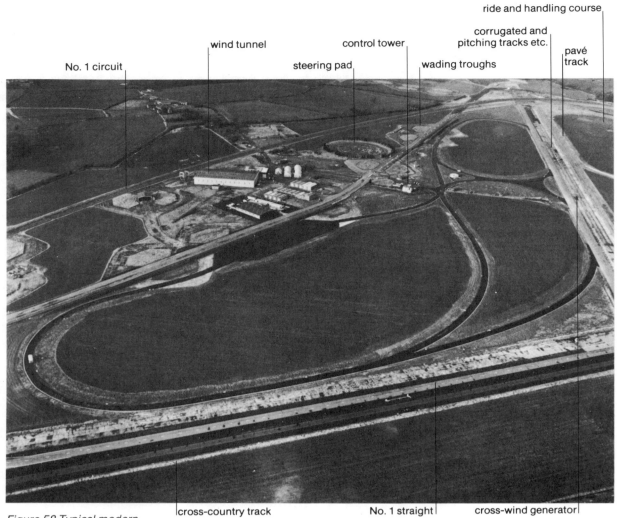

No. 1 circuit   wind tunnel   steering pad   control tower   wading troughs   corrugated and pitching tracks etc.   ride and handling course   pavé track

cross-country track   No. 1 straight   cross-wind generator

Figure 58 Typical modern proving ground

66

REDUCING FUEL CONSUMPTION

5.1 Introduction

For decades futuristic science-fiction motor cars of various kinds have been doodled, drawn, sculpted and occasionally even built by automotive enthusiasts (Figure 59). The desire to define and describe the car of the future in this way was never more intense than in the fuel crisis of the early 1970s. For the first time since the birth of the motor vehicle, the world was forced to accept what had always been true but seldom before conceded: that the earth's oil wells must inevitably dry up, sooner or later. Alternatives to conventional petrol and diesel engines were therefore sought with some desperation, so that a number of highly unrealistic concepts were considered.

But the real car of the immediate future is likely to be a relatively sober and sensible machine, superficially quite close in appearance and construction to the car of today. For although hundreds of millions of pounds have been spent during the past twenty years on the development of alternative sources of power, none *so far* has proved a serious rival to the modern petrol and diesel engines.

In addition, the pressure to take risks on these unproven alternatives is now less than it used to be: the balance between the world's consumption of energy and its production is today quite different from what it was in the early 1970s. In those days all kinds of fossil fuel, but particularly crude oil, were consumed with an extravagant disregard for the future and at a rate that increased steadily at about 3–4 per cent annually. Only by discovering a major new oilfield every year – a highly unlikely occurrence – could such a level of demand have been satisfied.

Since then, however, things have changed, partly as a result of the economic recession but partly through the more efficient use of energy in almost every industrial and domestic application, not least the motor car. The demand for all types of energy has not grown as dramatically as once expected, but has stabilized at a level that is expected to remain little changed to the end of this century.

More important, though, is the realization that not only are there reasonable grounds for inferring the existence of considerable additional reserves, but also that there are huge reserves of other types of fossil fuel, including coal, oil shale (a soft, slatey oil-bearing rock) and tar sands. These add up to a fuel supply that in total should last for anything from 240 years to 2200 years (see Figure 60).

Figure 59 Spectacular but impractical was the character of most design studies and styling exercises until quite recently

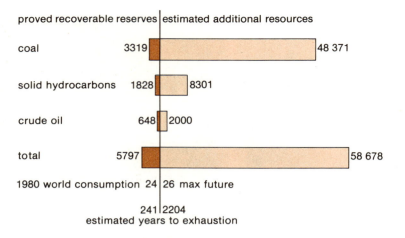

Figure 60 Energy scenario

Figure 60 Energy scenario

There is no immediate risk of the earth's fossil fuels running out if coal and tars are taken into account (Source: Ford Energy Report)

This scenario doesn't take into account the possible development of 'clean' nuclear fusion power, or the possible increased availability of renewable sources of energy such as hydroelectric, wind and wave power.

Although fossil fuel may remain relatively accessible for a long time to come, it certainly isn't going to remain cheap. It will be necessary, for instance, to drill for crude oil in ever more remote, inaccessible and inhospitable places. Already North Sea oil and Alaskan oil are much more expensive to obtain than Middle East oil, while the processes required to extract oil from shale or tar sands are vastly more expensive again. Yet at the same time the motorist of today isn't prepared to make more than a modest sacrifice in performance (cruising ability, acceleration and so on) in return for improved fuel consumption.

If we assume no sudden change in users' demands, nor in the large-scale production methods of manufacturers, the requirements for the next generation of cars are fairly clear:

They must be similar in performance to present-day cars.

They must have significantly better fuel consumption.

They must use technology that is not greatly different from that of today.

Even with *modest* extensions of present technology, the improvement in fuel consumption is likely to be very great. Already Renault is expressing confidence in its ability to meet an overall fuel consumption target of 94.2 m.p.g. (3 litres/100 km) for their Vesta. This target is for a complete driving cycle including some urban driving. The Vesta (Figure 61) is an experimental economy car that Renault is developing with 50 per cent financial backing from the French Government.

These figures for fuel consumption might seem to you in the realm of fantasy, but we shall see that a plausible series of technical changes can get close to this low level of consumption.

To put it as simply as possible, there are three main ways in which a car can be made more fuel-efficient:

by improving the efficiency of the engine in extracting energy from the fuel;

by matching the rotational speed of the engine more effectively to road speed;

by making the car easier to push along.

There are combinations of ways of achieving these targets. For example, a car can be made easier to push along if it is lighter and its shape does not build up air resistance. In section 6 I shall deal with ways of achieving aerodynamic efficiency, but here I am going to look at the mid-term targets and the industry's options for improving fuel consumption. The main focus of our attention is improved engine burning, improved transmission and weight reduction.

Figure 61 Renault Vesta, an experimental economy car, in clay-model form. Athough only 10 ft 6 in long, the finished car is expected to have a drag coefficient of 0.25

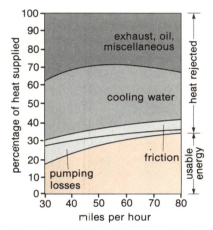

Figure 62 Division of energy within a petrol engine, giving an overall efficiency of about 30 per cent

5.2 The petrol engine

To understand how the new levels of fuel consumption might be attained it is best to begin with some basic facts about the conventional petrol engine.

The conventional four-stroke petrol engine is at best only about 30 per cent efficient. About a third of the energy from the petrol is dissipated in exhaust gases, about another third is lost in heat through the cylinder walls and engine block. The remaining third is available as usable power, although, as we shall see, much of that power never reaches the wheels (Figure 62).

Various attempts are being made to attack the basic inefficiencies. For example, turbocharging recycles exhaust gases and is able to extract a further reserve of power from what would otherwise be wasted.

Other lines of attack go directly to the fuel mixture itself. The mixture of petrol vapour and air sucked into the engine has to strike a balance between an over-rich mixture, which is not completely burnt, and therefore wasted, and an over-lean mixture, which cannot go below a certain limit without loss of power and misfiring.

A further major limitation of the petrol, or spark-ignition, engine springs from the fact that its efficiency is much influenced by its *compression ratio* (Figure 63): the ratio of the volume above the piston at bottom dead centre to that above it at top dead centre. Efficiency increases with higher compression ratios, but the highest usable compression ratio in a petrol engine is limited by the onset of an unwanted phenomenon called detonation, knock or pinking (after the noise it makes). When a fuel detonates, it explodes violently, uselessly and damagingly rather than burning smoothly and relatively slowly.

Before we go any further into engine design, let's consider the fuel itself. Read the technical box on petrol and its additives and then answer SAQ 12.

Figure 63 Compression ratio =

$$\frac{clearance\ volume\ +\ swept\ volume}{clearance\ volume}$$

Petrol vapour in the cylinder is compressed to a fraction of its original volume

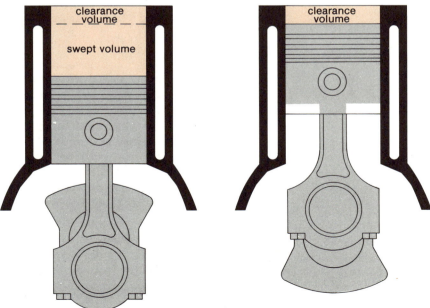

69

Petrol and its additives

Petrol is the refined product of naturally occuring oil deposits (petroleum). These deposits are organic in origin. They occur in sedimentary rocks and are thought to derive from marine algae and related life forms (unlike coal, which derives from plants). Petroleum is almost exclusively composed of compounds of carbon and hydrogen (hydrocarbons).

It is calculated (1982) that the world reserves of crude oil amount to 650×10^9 barrels (1 barrel = 42 gallons).

As a motor spirit, petrol needs to have the following characteristics:
high volatility to permit spark ignition,
ability to fire in cold weather,
no vapour lock in hot weather,
smooth fuel burn without detonation (knocking).

The tendency to detonate is influenced by a number of factors, one of which is combustion-chamber design, but for a given sort of combustion chamber, the detonation tendency is in turn limited by a special quality of petrol measured by its *octane rating*: the higher the octane rating, the higher the usable compression ratio and the higher the efficiency.

One way of producing fuel with a high octane rating is simply to formulate it from the lighter, more highly refined constituents of crude oil. But the more highly refined the fuel, the more expensive it becomes, and in two ways: it costs more to extract and constitutes a smaller proportion of each barrel of crude. It has thus been a common practice for the past fifty years to formulate petrol from moderately refined crude oil, the octane rating of which is raised by a special additive discovered in the 1930s: tetraethyl lead.

For decades this practice went unchallenged, but it involves the release of large amounts of lead, which is poisonous, into the atmosphere every year. The effect of this on health, particularly of children, has caused increasing concern in recent years. As a result, petrol lead levels have been reduced in most countries: from 0.64 g/litre in 1971 to 0.15 g/litre in 1976 for West Germany, for example, and from 0.84 g/litre in 1972 to 0.45 g/litre in Great Britain, with legislation enacted to bring the level down to 0.15 g/litre by 1985. Lead-free petrol has been available in the United States of America for some years, but not as a measure to reduce environmental lead levels themselves, but because lead spoils the filtering catalysts used to reduce exhaust pollution of other kinds.

In Britain, pressure is growing either to eliminate lead from petrol altogether or at least to introduce a lead-free grade as soon as possible. This move is opposed by some automotive engineers on the general grounds that it would reverse, even if only temporarily, the increasing tendency to use the high-compression, high-efficiency engines that are now so badly needed to cut fuel costs. The oil companies claim that refining costs would be increased very substantially, for a 97 octane no-lead grade, for example. There is the added complication that existing engines would tend to suffer valve damage without at least a small amount of lead in the fuel they use. A sensible compromise, perhaps, might be to introduce in two or three years' time (giving the motor manufacturers time to make the necessary design changes) a lead-free 92 octane fuel for new models, and at the same time to retain for some years a low-lead 97 octane fuel, which could also be produced at modest additional expense, for existing engines. Clearly, though, the conservation of energy conflicts in this case with the improvement of the environment in a rather complex way. But it does seem likely that improvements in engine design would very soon compensate for any deterioration in fuel consumption brought about by the introduction of a lead-free 92 octane fuel.

Petrol-engine design trends

The optimistic view I've presented of fuel efficiency is based on a number of trends. The first is that through a number of individually small but collectively significant modifications the petrol engine has improved noticeably in efficiency since the first fuel crisis of 1973. Many engines designed to run on 97 octane fuel now have compression ratios around 9:1, while for a few the figure is as high as 10:1. These changes may be far reaching in the design of other components, as the following example shows.

Raising the compression ratio of a petrol engine will result in an increase in the thermal efficiency of the engine. An increase in the efficiency of an engine means that more power can be obtained from the same amount of fuel burnt. This achieves fuel economy, which is the desired end result. But the effect of changing the compression ratio has undesirable effects on other components. High compression ratios generate more forces on the pistons, wrist pin, connecting rod and crankshaft. These parts have to be made from stronger materials or to have larger cross-sections in order to sustain the higher forces. Design changes in these components may lead to heavier structures, the use of different materials and, perhaps, other production methods.

The use of heavy structures leads to a heavier engine, which results in an increase in the total weight of the vehicle and a change in weight distribution. The increase in vehicle weight requires more power from the engine for propulsion, and consequently more fuel to produce this power. This may offset the saving in fuel earned initially. Changes in weight distribution between front and rear may alter the vehicle's handling characteristics.

The use of stronger materials and different manufacturing processes will certainly affect the production cost of the engine.

Changes in the pattern of forces of the modified engine will influence the design of engine mounting, clutch and transmission. Additional changes may be required in the combustion chamber and piston head to improve combustion. Thus the effects of raising the compression ratio in the engine could lead to changes all the way from the piston head to the rear wheels. What might seem to be a simple change has extensive repercussions.

Apart from raising the compression ratio another avenue of progress concerns a special sort of combustion chamber, which in the first place generates plenty of turbulence, or violent local movements in the incoming air and fuel. This speeds up the combustion so that very economical lean mixtures can be used. At the same time it must be designed to allow the use of very high compression ratios without inducing detonation when running on the richer mixtures needed for full power. These conflicting rquirements seem to have been reconciled in the 'Fireball' combustion chamber invented by the Swiss engineer Michael May and adopted by Jaguar for their V12 engines. It incorporates an inlet valve lying in a shallow recess, which is connected by a channel to a deeper bathtub-shaped pocket surrounding the exhaust valve. The upward movement of the piston creates high-velocity swirl and a looping flow pattern that concentrates the charge near the exhaust valve (Figure 64). With Fireball combustion chambers the Jaguar XJS and V12 models gained a slight increase in power, a significant reduction in exhaust gas pollution and a fuel consumption improvement in the region of 20 per cent. Other manufacturers are likely to introduce similar lean-burning high-compression

71

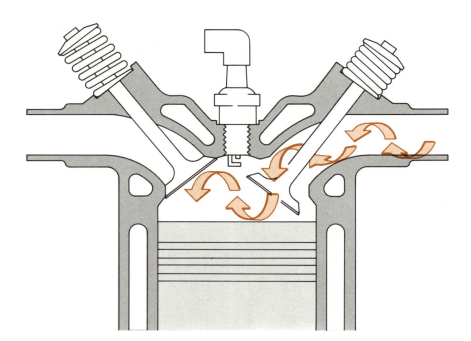

Figure 64 Combustion of lean fuel–air mixtures is improved by giving the gases a swirling motion

combustion systems during the next few years. All of them will depend critically on the complex and sophisticated electronic engine management systems that are gradually being adopted for existing engines, which provide precise control of ignition timing, mixture strength and other factors.

A further cause for much attention is exhaust pollution control, which is dealt with in section 8.5. For the moment it is sufficient to say that legal controls seek to limit three substances that occur in automotive exhaust gases: unburnt hydrocarbons, oxides of nitrogen and carbon monoxide. There is much debate about the toxicity of these substances. No one argues that they are good for you, but there is some uncertainty about what level it is reasonable to accept in urban areas. The lead levels of inner London are seventeen times higher than those of the suburban fringes.

What is more certain is that the overall effect of this exhaust pollution control has not been as wholly beneficial as was perhaps at first hoped. One problem is that the oxides of nitrogen, which the more demanding sets of regulations seek to limit, tend to be produced at precisely those high combustion temperatures that are conducive to greater efficiency. Strict exhaust pollution control, therefore, as enforced in California and Tokyo, not only increases fuel consumption, and thus is in direct conflict with energy conservation, but also reduces the performance of a car and adds very considerably to the cost and complication of the engine.

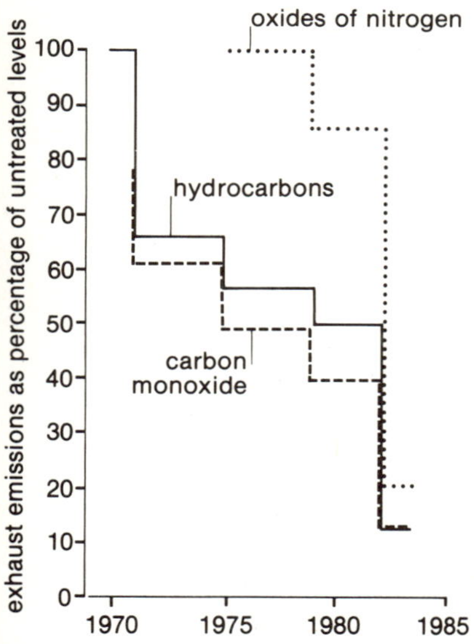

5.3 The diesel engine

Some, but not all, of these conflicts can be avoided by the use of the diesel engine. This is considered by some engineers to be the power unit of the future. The first of the diesel engine's several fundamental virtues is that it does not suffer from detonation and so can run at very high compression ratios, ranging from 16:1 to 23:1, to attain excellent efficiency. Indeed, these high compression ratios are needed to raise the temperature of the incoming air to the point at which it will ignite the fuel, for there are no spark plugs and the fuel is squirted into the combustion chambers under very high pressure through special injectors.

Two further reasons for high efficiency follow from the 'stratified-charge' principle, upon which all diesels depend. In a stratified-charge engine the fuel and air are persuaded to form a central kernel of fuel that is always rich enough to burn, no matter how small the dose of fuel or large the amount of

Figure 65 Reductions in exhaust emission as required by European legislation (Source: Downs, 1978, Figure 2b)

air, a kernel surrounded by strata, or layers, of progressively decreasing richness, to which the flame spreads. Thanks to charge stratification, a diesel's power output can be controlled by varying the supply of fuel alone, not of fuel and air together. Hence the engine can be allowed to induce air unchecked, so a throttle valve is not needed and pumping losses can be reduced. In addition, very weak mixtures can be used under part-load conditions, thus permitting significant fuel savings and leading further to high efficiency. As a result of this efficient combustion, emissions of hydrocarbons and carbon monoxide are very low, but a diesel cannot easily be made to meet very strict limits on the emission of oxides of nitrogen.

One proposal for a new form of diesel engine is that it should be completely insulated, without any water coolant, water pump or radiator. Such an 'adiabatic' engine would operate at higher temperatures and all internal surfaces (such as cylinder bores, piston crowns, valve faces and exhaust parts) would have a ceramic coating to prevent heat losses to the body of the engine. 'Adiabatic' means change of pressure without loss of heat.

The gains in engine efficiency can be expressed as amount of fuel consumed per horsepower developed over one hour's test run. For comparison, typical figures are:

petrol engine	0.5 lb/hp/h,
diesel engine	0.4 lb/hp/h,
improved diesel engine	0.35 lb/hp/h,
adiabatic engine	0.28 lb/hp/h.

Were such an 'adiabatic' engine developed commercially, it could offer substantial savings in fuel.

Unfortunately, though, the diesel has its drawbacks. For example, diesel engines are generally a good deal heavier than petrol engines of the same capacity and invariably are much less powerful. For a long time this meant that diesel-engined cars were so slow and sluggish that they offered only a marginal advantage in fuel consumption over what might be obtained from a petrol-engined car detuned to the same level of performance. This situation is being changed by the introduction of new lightweight high-speed diesels, such as the engine developed by Volkswagen for the Golf.

The diesel engine's performance is being further boosted by the increasingly widespread use of turbochargers. A turbocharger consists of a turbine driven at high speed by the exhaust gases (thus utilizing energy that would otherwise be wasted) coupled to a centrifugal compressor, which crams more air into the cylinders than they could draw in unaided. With more air available, more fuel can be burnt and hence more power developed.

The need to add a turbocharger, however, highlights what is perhaps the diesel engine's greatest disadvantage: its cost and complication. Even without a turbocharger, a diesel engine must be stronger than a petrol engine of the same power, to withstand the higher combustion pressures, and it needs a more robust transmission system as well as heavier auxiliary components. However, the biggest cost penalty comes from the very expensive, high-precision, high-pressure fuel-injection pump it needs, the price of which alone can be about half that of a complete small petrol engine. Add a turbocharger, and the result is an installation far more costly than a conventional petrol engine with a carburettor.

Diesel cars are therefore a good deal more expensive to buy than their petrol-engined equivalents, costing anything from about £600 to nearly £3000 more at 1982 price levels. This means that in countries where the price of diesel fuel is similar to that of petrol, there's little to be gained by purchasing a diesel car; it has to be run for anything from 60 000 to 100 000 miles before the savings, due to its superior fuel consumption, compensate for its extra initial cost.

It could be argued, though, that if more Governments agreed to reduce the taxes levied on diesel fuel, significant long-term savings would result. Apart from the undoubted consumption advantage of a diesel engine, the fuel it uses is a little cheaper to refine than petrol and it needs no lead additive.

The future of the diesel

Furthermore, there's considerable potential for further improvement in diesel engines for motor cars. This is because cars use the indirect-injection type of diesel engine (Figure 66a), in which the initial kernel of rich mixture is formed within a small auxiliary combustion chamber communicating with the main one. However, the movement of gases between the two chambers makes it less efficient than the direct-injection type of diesel. In an engine of this kind (Figure 66b) a tangential inlet port, sometimes working in conjunction with a masked inlet valve, imparts to the fresh charge of air a violent swirling motion, which is accentuated by the upward movement of a bowl-in-piston combustion chamber. The fuel is injected directly into this combustion chamber and the swirling action creates the necessary kernel of rich mixture with its surrounding layers of progressively decreasing richness. Such engines have been commonplace in big trucks for many years, but until quite recently have been incapable of exceeding about 3500 r.p.m. or so. However, high-speed direct-injection diesels suitable for cars are being developed.

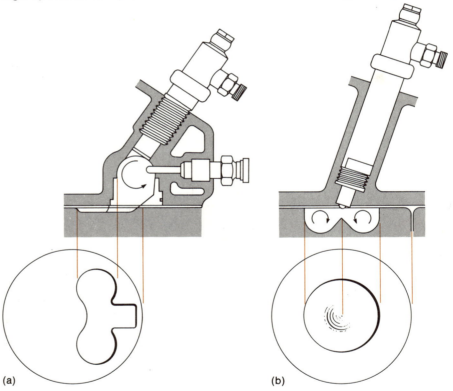

Figure 66 Two types of fuel injection in diesel engines

*(a) Indirect injection: the fuel is injected into a small 'pre-chamber' connected by a passage to the main combustion chamber below, which is much larger (except when the piston is at top dead centre). The horizontal device protruding into the pre-chamber is a glow plug, which provides additional heat for starting
(b) Direct injection, with a bowl-in-piston combustion chamber*

(a)   (b)

SAQ 13

What are the essential differences between a petrol engine and a diesel engine?

5.4 Matching the engine to the car

The improvements to diesel engines just described would be expensive to make. So engineers all over the world are looking for cheaper ways of improving fuel consumption, and one possibility, which would be of great benefit to the ordinary petrol engine, but would be advantageous for the diesel too, is an advanced form of transmission system.

The problem is that to provide acceptable acceleration and performance, existing transmission systems do not allow the engine to run for long at the speeds that give the best fuel consumption for any power output required. What is needed, therefore, is a transmission system that allows the engine to run at the very low rotational speeds and large throttle openings conducive to efficiency, except when moving off from rest, when accelerating hard or when cruising at very high speed. Under these latter conditions the engine must be allowed to run at higher speeds to develop the higher levels of power required.

Such a transmission, designed for economy, would need to be fully automatic, with a sophisticated electronic control system to keep the engine in its most efficient mode of operation for the power output required. Unfortunately, conventional automatic transmissions involving an epicyclic gear train and a hydrokinetic torque converter are not suited to this purpose, because the fluid-churning frictional losses in the torque converter rule out any hope of obtaining improvements in fuel consumption.

To avoid these losses, a number of manufacturers are considering for their next generation of cars the possibility of modifying existing manual gearboxes, which are mechanically very efficient, to work fully automatically. Such a gearbox would need at least six speeds to give the spread of ratios required, and would work in conjuction with friction clutches: a main clutch for moving off and an auxiliary clutch for gear changes on the move. The British component manufacturer, Automotive Products, has experimentally developed a gearbox of this kind (Figure 67).

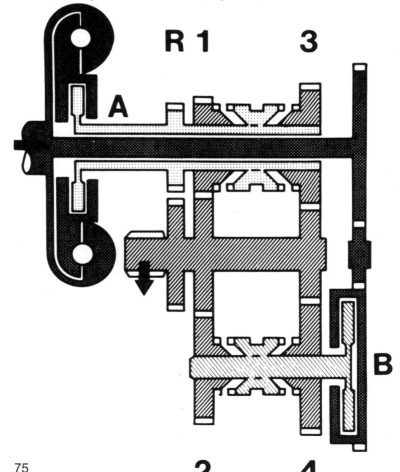

Figure 67 Four-speed three-shaft version of the experimental automatic gearbox developed by Automotive Products

While second and fourth gears are driven through a normal clutch, a secondary 'wet' clutch is added, through which reverse, first and third gears are driven. Both gear selection and clutch engagement are hydraulically actuated and electronically controlled, including the progressive engagement of the main clutch when moving away from rest. A six-speed version could allow significant fuel savings

Long-term hopes for the future rest on the development of an efficient form of continuously variable transmission. One concept that is being extensively explored is a transmission using a belt and expanding pulley of the kind used for many years in Daf and small Volvo cars. About the same level of frictional loss as in a conventional automatic transmission is associated with the rubber belts used, but the original creators of the system have developed a completely new version using more stable belts of linked steel blocks. These belts are much more efficient than their predecessors and the transmission as a whole more compact. Figure 68 illustrates the principle of this form of continuously variable transmission.

At bottom right is the input from the engine to the transmission through a more or less conventional clutch. The power is transmitted to a rotating drum, which is in two halves, one of which can move laterally. Each half-drum, or sheave, has a conical face. Thus, when the two halves are pushed together, the belt is forced to rotate around the largest diameter; when the drums are drawn apart, the belt drops down to a smaller diameter. Similarly, at the output end there is another split drum acting as an expanding and contracting pulley. Sensors change the shape of the vee-pulleys relative to the belt, in this way changing their effective diameters. The gearing is continuous across the range provided by the set of pulleys.

Van Doorne of Holland originally developed this form of transmission, calling it 'Transmatic'. Fiat and Borg Warner are promoting it vigorously. It is very suitable for front-wheel-drive transverse-engine cars, giving a gear ratio of about 6:1.

Another promising form of continuously variable transmission is the Perbury type, involving rotating discs with toroidal surfaces, between which are clamped sets of tiltable rollers (Figure 70). Such a system promises to be highly efficient. It has a good ratio spread and can be coupled to an epicyclic gear train to become a compact two-range system with a geared neutral and a reverse.

With new systems of transmission based upon these well-advanced prototypes the improvements in fuel efficiency are likely to be of the order of 20 per cent, and that figure does not take into account other means of conserving fuel.

5.5 Reducing resistance to motion

So far we have considered ways of improving the fuel consumption of the motor car by increasing the efficiency of its engine and by matching the rotational speed of the engine more effectively to road speed. There is another, quite different way of improving a car's fuel consumption, by making it easier to push along.

This can be achieved in three main ways:
by reducing the air resistance to the body shape,
by reducing the total weight,
by minimizing tyre rolling resistances.

The cheapest way of reducing a car's resistance to motion is to reduce its aerodynamic drag or wind resistance. Reducing aerodynamic drag by 10 per cent, for example, improves fuel consumption by about 3 per cent. Thus, by reducing the drag coefficient of a car from 0.45, the present average for European cars, to 0.25, which is the lowest level currently thought practicable, its fuel consumption could be improved by over 12 per cent.

The ways in which these gains can be realized in practice are described in section 6. It's worth noting, though, that several cars introduced towards the end of 1982 exemplified the strong trend towards lower drag coefficients for cars: 0.30–0.32, depending on the version, for the Audi 100; 0.32–0.34 for the Ford Sierra; 0.35 for the Citroën BX and 0.38 for the Opel Corsa.

Figure 68 Van Doorne Transmatic
continuously variable transmission

Figure 69 V-belt automatic
transmission on a small formula
racing car

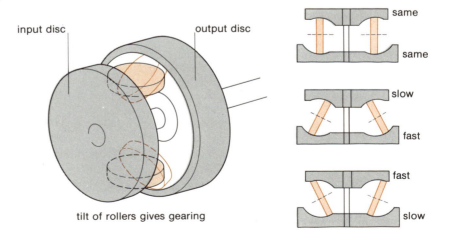

Figure 70 Perbury type of
continuously variable transmission

Tilting rollers between the driving
disc and the driven disc allow the
effective gearing ratio to be
continuously varied

Weight and new materials

The weight of a motor car has a profound influence on its fuel consumption. Every time a car is accelerated or made to climb a hill, a force proportional to its weight (or strictly speaking its mass) has to be overcome. Indeed, it is known that every 10 per cent reduction in weight means a reduction in overall fuel consumption of about 4 per cent, if engine and transmission are tuned to take account of it. The major components of mass-production cars are designed by weight watchers, but the potential for weight reduction in European cars of small or medium size is so far relatively small. As explained in section 4, safety requirements put limits on the minimum weight that can be achieved. Already most body shell structures have become highly efficient and light in weight with the help of computer-aided design techniques introduced in the 1970s.

Table 4 shows the target weights of the Ford experimental mid-engined Mustang (not the production Mustang mentioned in section 4.2), giving a total of nearly 2000 lb, one quarter of which is the weight of driver, passenger, luggage and fuel.

Major manufacturers are aiming for a 10 per cent reduction in weight over these kind of figures. In other words, the weight saving is to be roughly half the weight of the engine.

Table 4 Target weights for experimental Ford mid-engined Mustang

Name	Total weight/lb	Front axle load/lb		Rear axle load/lb	
body	466	198		268	
frame	132	63		69	
engine	270	44		226	
front suspension	54	54		0	
rear suspension	78	75		3	
brakes	80	46		34	
transaxle and clutch	121	−4		125	
wheels and tyres	90	57		33	
exhaust system	16	−1		17	
fuel system	28	34		−6	
steering system	18	13		5	
cooling system	19	4		15	
electrical system	59	42		17	
oils	17	3		14	
shipping weight	1448	628	43.4%	820	56.6%
gasoline	78	90		−12	
water	18	4		14	
curb weight	1544	722	46.8%	822	53.2%
driver only					
full tank	1719	807	46.9%	912	53.1%
empty tank	1641	717	43.6%	924	56.4%
driver and passenger	1894	892	47.0%	1002	53.0%
driver, passenger and 50 lb *luggage*	1944	888	45.6%	1056	54.4%

Source: Korff (1980), Table 11, p. 205.

How might this be achieved? Obviously by changing to lighter materials. I have already talked a little about aluminium in relation to the body shell. Aluminium, unfortunately, is not only several times more expensive than steel, but likely to become even more expensive as time goes on because its extraction consumes large amounts of electricity.

While its use for cylinder blocks, axle casings, etc. is nevertheless likely to continue or even increase, it is unlikely to be adopted as the basic material for a car's body shell, although it may provide a body frame, as in BL's research vehicle ECV3.

Modern plastics, however, offer more hope for weight reduction. They are corrosion-free, can readily be formed into complex shapes and can generally be produced at a lower energy cost than steel. Even if they were produced at a slightly higher energy cost, their use would still lead to significant overall savings, since a car typically consumes ten times more energy during its operating life than it requires for its initial creation. Today's cars therefore incorporate quite large amounts of plastics. The new Citroën BX in Figure 71 is a good example: its bonnet and tailgate frame are made of plastics as well as its bumpers and a number of other smaller components.

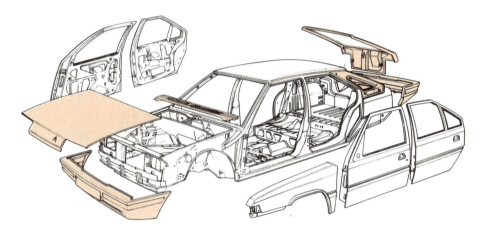

Figure 71 Citroën BX
The shaded areas show the extensive use of plastics in the external body work

Difficulties arise, though, if plastics are used as structural materials for the actual frame or body shell of a car. Most types are not stiff enough and so need reinforcement with fibres of glass or some other material. They also tend to soften under the high temperatures encountered during the body painting process and even in hot climates. Large fibre-reinforced body panels or structural elements take longer to manufacture than steel equivalents. A steel body press takes seconds, any plastics process takes minutes or hours. Glass-fibre reinforcement in plastics has given way to carbon-fibre reinforcement as the most fruitful field of research. As early as 1977 Ford produced a full-size car with a carbon-fibre plastics body comparable to a standard metal-built car. The weight saving was a startling 33 per cent.

Table 5 shows how a small European car could benefit from plastics substitutes. Because plastics are derived from petrochemicals it is thought that using plastics is unwise and wasteful in energy terms. This is not necessarily the case. Table 6 shows the energy content of basic materials. Steel absorbs $2\frac{1}{2}$ times as much energy in production (extraction, refining, rolling and supplying) as supplying the same weight of plastics. More than that, in many applications only half the weight of plastics is needed to do the same job as steel.

A 10 per cent saving in weight is worth while when you recall that it gives a saving of about 4 per cent in fuel consumption, or a saving of 600 litres over the car's lifetime, assuming 150 000 km travel.

Table 5 Weight savings by using graphite-fibre-reinforced plastics in a small car

Component	Weight/lb		
	in steel	in plastics	savings
body-in-white*	461.0	208.0	253.0
frame	282.8	207.2	75.6
front end	96.0	29.3	66.7
hood	49.0	16.7	32.3
deck lid	42.8	13.9	28.9
bumpers	123.1	44.4	78.7
wheels	92.0	49.3	42.7
doors	155.6	61.1	94.5
bracketry, seat frame, etc.	69.3	35.8	33.5

*Complete body structure without doors, deck lid, etc. and trim.
Source: *Ford Energy Report*, p.117.

Table 6 Energy content of basic materials

Material	Energy content/ (kJ per cm^3)
aluminium	550
steel	380
nylon	270
polyurethane	100
polypropylene	70

Source: *Ford Energy Report*, p.117.

Tyre rolling resistance

While most of the resistance to motion experienced by a car is created by aerodynamic drag, a significant amount is due to friction in the carcass bracing cords and treads of the tyres. The resistance of the tyre depends not only on its pressure but also its temperature. A cold tyre will have double the resistance it has when 'warmed up' after 50 miles. The losses at the tyres are almost entirely in the flexion or hysteresis of the tyre, with a small loss going to surface friction, and local aerodynamic turbulence at the wheels.

For a typical urban driving cycle at slow speeds, 20 per cent of the driven axle energy is lost to the tyres and 27 per cent to aerodynamic drag. However, at motorway driving speeds the proportions change and tyres account for about 15 per cent or less.

Improvements in efficiency are not easily won: too stiff a carcass leads to riding discomfort, while good wet grip depends largely on the very hysteresis losses in the tread rubber that increase rolling resistance. Nevertheless, new tyres of low rolling resistance have been introduced by all the major manufacturers, while Dunlop, in conjunction with Shell, have developed a new material called Cariflex, which reduces energy losses at little sacrifice in wet grip.

Typical cross-sections of the new forms of tyres are shown in Figure 72. Radials are estimated to give an improvement of 5 per cent in overall fuel efficiency. The TRX types are even more efficient.

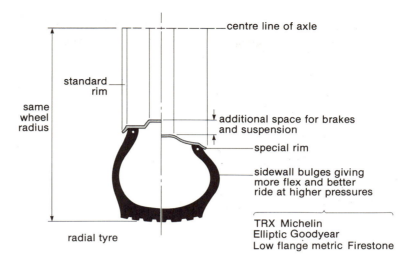

centre line of axle

standard rim

same wheel radius

additional space for brakes and suspension

special rim

sidewall bulges giving more flex and better ride at higher pressures

radial tyre

TRX Michelin
Elliptic Goodyear
Low flange metric Firestone

Figure 72 Tyre cross-sections compared

In this section I have laid out the likely improvements in engine design, in transmission and in rolling resistance. In total these gains in efficiency, as measured by fuel consumption, would be around 45 per cent. But that is not the end of the line. Another area of improvement, which is easier to facilitate, will be discussed in section 6.

SAQ 14

Apart from improvements in engine design, what are the main ways of reducing the fuel consumption of a typical car?

SAQ 15

What are the advantages of plastics parts in cars?

6 THE SHAPE OF THE FUTURE

In this section I shall describe some of the work of pioneers in aerodynamic cars, who had arrived at a clear consensus by the mid 1920s.

The general principles of aerodynamics are also discussed and some of the theoretical and modelling difficulties explained. Despite the theoretical uncertainties and the difficulties of modelling air flow, researchers have developed a clear understanding of the optimum low-drag form, the ideal shape.

However, even if ideal optimum shapes can be defined and made, they are just not very practical when applied to passenger-carrying vehicles. Not only that, the form of cars has responded in the past primarily to styling, safety and manufacturing. This means that for mass-produced cars aerodynamics have been neglected ('shamefully neglected' is Tony Curtis's view). Later in this section I shall try to describe the principal inhibitions on aerodynamic design.

Nevertheless, the advantages in fuel saving of low-drag cars can be clearly demonstrated and, belatedly, the car industry is taking up aerodynamics as a cause. But aerodynamic principles *can* be in conflict with the demands of the market place. To put it at its most brutal, it is no use designing an optimally efficient shape if no one will buy it. The conventional wisdom of several generations of designers and market researchers was that 'people won't buy cars shaped like potatoes' (quoted in Korff, 1980, p.34).

The television programme 'The shape of cars to come' explores the tension between the rational principles of aerodynamics and the more obscure market forces. It will help you to appreciate the arguments in the television programme if you have read this section before watching the programme. If you have not had an opportunity beforehand, read it as soon as you can after watching. This section gives you most of the technical information; the television programme is mainly about difficulties of integrating technical requirements with less clearly defined objectives.

6.1 The time tunnel

The earliest experiments in wind tunnels concerned aircraft, and date at least from 1871 in the United Kingdom and the work of Wenham and Browning on wing shapes. In 1901 the Wright brothers conducted tests in a tunnel they designed and built for themselves in Dayton, Ohio. Parallel attempts at streamlining automobiles were made as early as 1899 and later 'ideal' shapes were based on the airship, torpedo or boat (Figure 73).

Figure 73 Early design for low-drag ground vehicle, around 1911

Alfa Romeo Castagna, with body designed for Count Ricotti

Figure 74 Edmund Rumpler's original Tropfenwagen of 1921

The open-topped version had a drag coefficient of 0.54

A continual fertilization of car design by aircraft engineers is the main feature of these early beginnings. After the First World War, Germany had a surplus of aircraft engineers and engineering facilities and no aircraft. The paradoxical result was that Germany established a ten-year head-start in the study of air flow around cars. The first automobile to be truly streamlined was introduced by Edmund Rumpler in 1921. His *Tropfenwagen* (teardrop car), shown in Figure 74, was conceived as a vertical aerofoil. A model tested in 1922 supposedly had a drag coefficient of 0.54.

However, Rumpler's approach was intuitive and he treated air flow as a two-dimensional problem.

Paul Jaray, Hungarian by birth, is generally recognized as the first person to set automobile aerodynamics on a methodical basis. Jaray was chief designer at the Zeppelin Airship Works between 1914 and 1923. With his assistant, W. Klemperer, Jaray conducted a comprehensive series of tests on shapes at 1:10 scale (Figure 75).

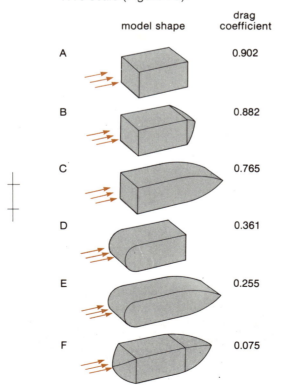

	model shape	drag coefficient
A		0.902
B		0.882
C		0.765
D		0.361
E		0.255
F		0.075

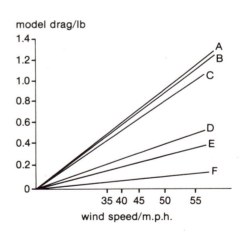

Figure 75 Examples of early wind-tunnel research by Paul Jaray and associates

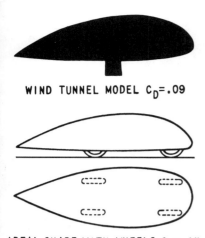

WIND TUNNEL MODEL $C_D = .09$

IDEAL SHAPE WITH WHEELS $C_D = .14$

Figure 76 Ideal shapes as derived by Paul Jaray, around 1920

Figure 77 Wind-tunnel model of Jaray compromise shape, reported to have a drag coefficient of 0.19

Figure 78 Volkswagen 'Beetle' prototype in wood, 1930

The history of car aerodynamics can be viewed as variations on the known *ideal shapes* determined by Jaray's work (Figure 76). He established the limiting drag coefficient for a ground vehicle with wheels, as being of the order of 0.14.

In 1922 the Zeppelin works built the first full-size Jaray body. The drag coefficient was shown to be about 0.30 compared to the usual 0.60 of that time. Jaray also devised an ultra-low-drag shape in model form, giving a coefficient of 0.19 (Figure 77). This shape is the clear forerunner of the Volkswagen 'Beetle' (see Figure 78). The form can be likened to two ideal aerofoil shapes one on top of the other.

As early as 1920 the aircraft designer Dornier had proposed a patent for a truncated, or blunt base, aerofoil. This idea was taken up by Wunibald Kamm, who showed it was possible to make low-drag shapes with a cut-off rear end. In these experimental cars, drag coefficients of about 0.23 were obtained (Figure 79). This kind of body shape bears the name of the originator (Kamm cars, Kamm tail, K-form).

By 1933 W.E. Lay, following the work of Jaray, Klemperer and Kamm, set up a well integrated series of wind-tunnel tests at the University of Michigan. He devised a set of interchangeable components and ran drag tests on all the permutations. Typical results are shown in Figure 80.

One of Lay's test series seems to be playing a practical joke on automobile stylists committed to elaborate streamlining. Lay showed that a rectangular box on wheels has a drag coefficient of 0.86, but by gradually rounding the corners, he achieved a value of 0.46, a great reduction, for relatively small physical changes. Later researchers observed that this Lay box has almost the same drag coefficient as the Volkswagen Microbus as measured in 1951 at the Braunschweig Polytechnic.

These early researchers not only investigated low-drag shapes in general, but also investigated the subsidiary problems that are still a major concern to contemporary investigators. These secondary problems include lift and pitching movement, side forces and yawing movement, internal air flow for cooling, aerodynamic noise, and the techniques of testing themselves (both on-road and off-road).

Figure 79 Kamm's experimental cars with truncated tail, 1940

The history of car aerodynamics can be viewed as variations on the known *ideal shapes* determined by Jaray's work (Figure 76). He established the limiting drag coefficient for a ground vehicle with wheels, as being of the order of 0.14.

In 1922 the Zeppelin works built the first full-size Jaray body. The drag coefficient was shown to be about 0.30 compared to the usual 0.60 of that time. Jaray also devised an ultra-low-drag shape in model form, giving a coefficient of 0.19 (Figure 77). This shape is the clear forerunner of the Volkswagen 'Beetle' (see Figure 78). The form can be likened to two ideal aerofoil shapes one on top of the other.

As early as 1920 the aircraft designer Dornier had proposed a patent for a truncated, or blunt base, aerofoil. This idea was taken up by Wunibald Kamm, who showed it was possible to make low-drag shapes with a cut-off rear end. In these experimental cars, drag coefficients of about 0.23 were obtained (Figure 79). This kind of body shape bears the name of the originator (Kamm cars, Kamm tail, K-form).

By 1933 W.E. Lay, following the work of Jaray, Klemperer and Kamm, set up a well integrated series of wind-tunnel tests at the University of Michigan. He devised a set of interchangeable components and ran drag tests on all the permutations. Typical results are shown in Figure 80.

One of Lay's test series seems to be playing a practical joke on automobile stylists committed to elaborate streamlining. Lay showed that a rectangular box on wheels has a drag coefficient of 0.86, but by gradually rounding the corners, he achieved a value of 0.46, a great reduction, for relatively small physical changes. Later researchers observed that this Lay box has almost the same drag coefficient as the Volkswagen Microbus as measured in 1951 at the Braunschweig Polytechnic.

These early researchers not only investigated low-drag shapes in general, but also investigated the subsidiary problems that are still a major concern to contemporary investigators. These secondary problems include lift and pitching movement, side forces and yawing movement, internal air flow for cooling, aerodynamic noise, and the techniques of testing themselves (both on-road and off-road).

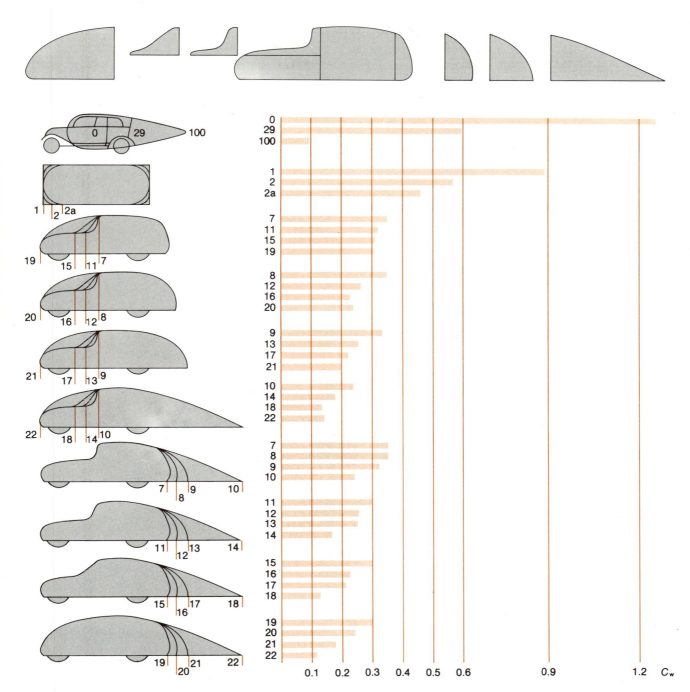

*Figure 80 Lay's wind-tunnel model
components and typical results*

SAQ 16

What was the contribution of Jaray to aerodynamic car design?

SAQ 17

What methods did Lay use in his aerodynamic tests?

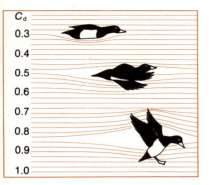

C_d	
0.3	
0.4	
0.5	
0.6	
0.7	
0.8	
0.9	
1.0	

Figure 81 Bird in flight

For gliding forward the drag coefficient is around 0.3. The drag increases to 0.5 when the wings flap and on landing to around 0.9

6.2 Principles of aerodynamics

As a car moves forwards it pushes against the air in its path, parts it, and sets up complex pressure variations and vortices in the vicinity. Certain shapes move through the air more easily than others. You can think of this by the analogy of pulling a block of wood through water by a piece of string. A square block creates more resistance than a sphere, and a boat shape or cigar shape causes less resistance still. Fish, for instance, are good aerodynamic (more accurately, hydrodynamic) shapes. Similarly, birds are also well formed aerodynamically, as you might expect (see Figure 81).

You may like to imagine that you have your hand sticking out of the window of a train moving at speed. The air feels almost solid and pushes against your hand with some force.

What factors make that force increase?

The main factors promoting air resistance are:

Firstly, the speed of the train; the faster the train goes, the greater is the force.

Secondly (this may not be obvious unless you have different size hands), a small hand will be resisted less than a large hand when held in the same position.

Thirdly, the way the hand is presented to the air flow; the hand held out flat experiences more resistance than a hand held sideways or clenched as a fist.

These three factors are illustrated in Figure 82(a).

Putting this in more general scientific terms, we can say that the total drag force F depends upon the velocity of the vehicle V, the size of the surfaces presented to the air flow A, and a coefficient C_d, the drag coefficient, that represents the slipperiness or cutting ability of the shape.

There are other factors that contribute to drag, namely the density of the air and its viscosity (or 'stickiness').

In fluid dynamics generally, viscosity and density can have a substantial effect on drag forces. However, when considering car shapes travelling through air, viscosity and density are relatively unimportant. Test results on aerodynamic drag can be corrected to that for air at a standard pressure and temperature.

Air conditions are not designed by car designers and in order to reduce air drag there are three main options for them:

reduce velocity V,

reduce frontal area A,

reduce drag coefficient C_d.

The velocity of the car is under the control of the driver, so only two factors remain to be manipulated by engineers and designers. The frontal area of ordinary passenger vehicles cannot go very much below about 1.5 square metres, so the main area of attack by designers is on the shape, or drag coefficient.

In rough terms it is easy to appreciate what is meant by a drag coefficient. A cubic block of wood generates a lot of resistance as it passes through the air and therefore has a high drag coefficient, which for a cube front on would be 1.10 (see Figure 82b). On the other hand an ideal cigar shape has a very low drag coefficient of 0.05 (low numbers mean low drag). Thus all motor cars

air velocity

size of frontal area

shape, drag coefficient

(a)

1.39	triangular rod
2.00	square rod
1.20	parachute
1.17	flat plate
1.10	cube
0.80	short cylinder
0.81	cube
0.51	cone 60°
0.41	hemisphere
0.34	cone 30°
0.08	wheeless teardrop
0.05	aerofoil

(b)

Figure 82 (a) Factors that contribute to air resistance
(b)Drag coefficients of simple shapes

have values somewhere between these two figures. The average drag coefficient for European mass-production cars in 1982 was about 0.45.

The drag coefficient is a convenient shorthand way of comparing the different shapes of vehicles. It is one important indication of the 'slipperiness' of the vehicle.

SAQ 18

What are the factors that contribute to air resistance?

SAQ 19

What does the term 'drag coefficient' mean?

6.3 Practical shapes

Whereas sections 6.1 and 6.2 dealt largely with historical research and the principles of aerodynamics here we look at aerodynamic practices in a less abstract way, as they operate in a world clouded by compromises and adjustment to other factors.

In theory an ideal shape for a ground vehicle might go as low as 0.1 in its drag coefficient. The Korff–Summers 'Goldenrod', which broke the world land speed record for a wheel-driven vehicle in 1965, reached a drag coefficient of 0.1165.

Yet the theoretical ideal shape, and such realized experimental forms, are not very realistic targets when placed against the requirements of a typical passenger car.

The ideal shape is over-long, cannot accomodate passengers easily, and has none of the details and protuberances characteristic of a modern car. In Figure 84 you can see the effect on the drag coefficient of gradually adding these features to the ideal shape. By adding a passenger volume and shortening the length we arrive at a drag coefficient of 0.26, then the wheel arches and wheels add 0.08, the underbody a further 0.05, the details of the upper body give an additional 0.04, and so on, until we reach a typical drag coefficient of 0.45, a more realistic figure.

In order to achieve a lower drag coefficient than this, a concerted revision of these characteristics is necessary. The state of the art of aerodynamics in 1982 is such that all major manufacturers are attacking these problems.

Figure 83 Record-breaking 'Goldenrod', designed by the Summers brothers and Walter Korff
This vehicle had a drag coefficient below 0.12

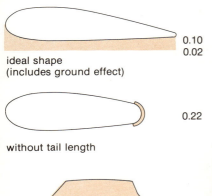

ideal shape
(includes ground effect)
0.10
0.02

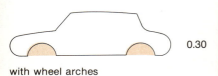

without tail length
0.22

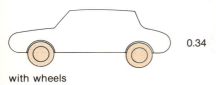

with passenger compartment
0.26

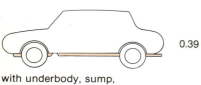

with wheel arches
0.30

with wheels
0.34

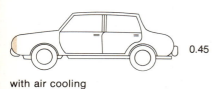

with underbody, sump,
exhaust, etc.
0.39

with trim, wipers, door handles,
window frames, etc.
0.43

with air cooling
0.45

*Figure 84 Build-up of drag from ideal
ground-vehicle shape to typical
saloon car*

In the television programme 'The shape of cars to come' the recently launched Ford Sierra is shown as an example of a new breed of aerodynamic car. The Sierra reduced the drag losses associated with upper-body details (such as wing mirrors and door handles), with wheels and with air intake. The drag coefficient of the Sierra is reported to be 0.34. Yet there is room for further improvements. Revisions to the wheel arches, the underbody panel and the details of window frames could reduce drag losses still further. The experimental vehicle Probe III developed by Ford *concurrently* with the Sierra shows some of these improvements in prototype form (Figures 85 and 86).

Figure 85 Ford Probe III, introduced at the Frankfurt Motor Show, 1981
This experimental vehicle is reported to have a drag coefficient of 0.22. Note the fared-in wheels and twin rear spoilers

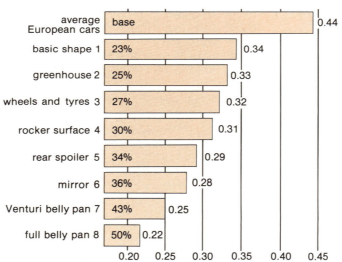

Figure 86 Step-by-step improvements of Probe III

In the development of the Probe III aerodynamic study vehicle a series of steps were developed to bring the drag coefficient down to 0.22. The effects of each step are shown here both as absolute numbers and as percentage improvement achieved (Results are from full-size wind-tunnel tests)

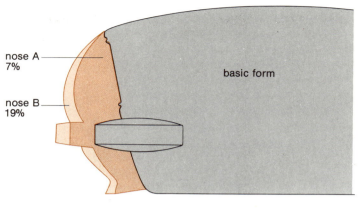

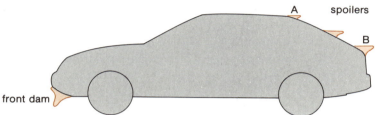

Figure 87 Add-on devices for reducing drag
Front dam and rear spoilers

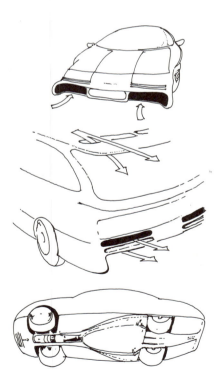

Figure 88 Further design ideas for reducing drag

Motor-car drag has been reduced by add-on devices such as the air dam (or under-bumper spoiler) and the rear spoiler. The main function of the front air dam, Figure 87, is to divert air from passing under the car between the underbody and the road surface. This achieves two objectives, one is to reduce drag by shielding the aerodynamic roughness of the underbody from the full velocity of the airstream, and the second is to reduce the pressure that might otherwise exist under the car. (Lift caused by pressure difference under and over the body profile has a significant effect on the directional stability of the car.) The effectiveness of the air dam in reducing drag varies from one car design to another but reductions in drag coefficient of up to 0.05 have been recorded as a result of adding an air dam.

The effect of rear spoilers in drag reduction arises from increased pressures on the rear facing body surfaces either downstream or upstream of the spoiler. On a 'squareback' design or one with a fairly upright 'hatchback' the spoiler is invariably located at the rear edge of the roof to ensure that the air flow separates approximately horizontally from this edge with no tendency to flow downwards over the back window. The arrangement increases the pressure acting on the back window area and, therefore reduces drag, but problems of dirt deposit increase.

For cars with fastback or hatchback shapes with a shallower back slope, the best location for a spoiler is at the lower edge of the inclined surface. It then serves to increase the pressure both upstream on the back window, and to a lesser extent downstream.

A similar spoiler location can be employed with a notchback body design, with similar benefits in drag reduction, especially if the boot upper surface is fairly sloping or its rear edge is well rounded. Shape and size of rear spoilers for maximum drag reduction are determined by wind-tunnel experiments. Reduction of as much as 0.04 in the drag coefficient, about 9 per cent of the drag of a typical car, can be achieved by adding a rear spoiler. Some further design ideas, as envisaged by Ford, for reducing drag are shown in Figure 88.

Once engineers are working at this level of refinement, factors beyond their control take on greater importance. For example, the improvements gained in re-designing the upper-body detailing (notionally 0.04) could be wiped out by the opening of side windows.

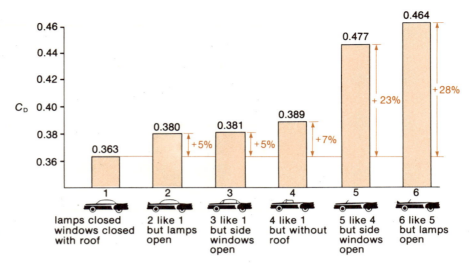

Figure 89 Effect of use on drag coefficient

Open windows or roof can substantially affect the overall drag of the car

Figure 89 shows the kind of reduction that can be predicted for different problems of use. The car shown is a low-drag type, comparable to the Ford Sierra.

Recent publicity might lead you to believe that drag coefficients below 0.45 are a very considerable achievement. The Austin–Morris Metro was widely acclaimed for reaching 0.41. However, you should not be too overwhelmed by this success. If we place such achievements in the perspective of recent history, then the Metro is seen as a modest improvement upon the Mini, roughly comparable with the Morris Minor and marginally worse than the Austin A40 (Figure 90).

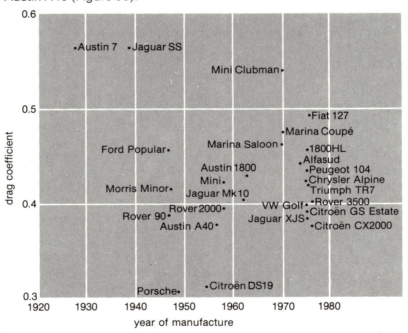

Figure 90 Drag coefficients of some well-known cars (Source: Engineering, August 1980, Figure 5, p.864)

Table 7 Drag coefficients of recent cars

Car	C_d
Escort	0.385
Metro	0.41
Renault 5	0.43
Fiesta	0.44
Polo	0.45
Fiat 127	0.46

The Metro no doubt is a more comfortable and better packaged car than its predecessors. The difference between the A40 and the Metro perhaps is rather more one of intention. The A40 is aerodynamically good by accident rather than by design.

However, other post-war cars were more deliberately aerodynamic and much more radical in their form. We can see that the Porsche 356 of 1949 and the Citroën DS19 of 1955 have drag coefficients just over 0.3. This is the background against which the drag coefficients of recent cars have to be viewed.

Exaggerated claims for improved aerodynamics should be viewed in two lights. Firstly, in light of what is theoretically possible, and secondly, in light of what has been achieved by the cars of the past in less fuel-conscious times.

SAQ 20

Which has the greater influence on creating drag resistance, the underbody components of the car, or the upper-body details (handles, wipers, wing mirrors, etc.)? Refer to Figure 84.

SAQ 21

The drag coefficient for the Metro is published as 0.41. How does this compare to its competitors and predecessors?

6.4 Benefits of low drag

The efforts that go into producing low-drag shapes would not make much sense unless they give specific *tangible benefits*. By now you will have grasped the general point that a low-drag shape means less air resistance, which in turn means less fuel to push the car through the air, but just how much fuel can be saved by the shape alone?

The answer to that is: it depends on the speed of the vehicle. The air resistance of any shape increases as the *square* of its velocity. Consequently, at a low speed the shape of the vehicle is relatively unimportant. For instance, at 20 m.p.h. the rolling friction of the tyres consumes more engine power than air resistance, but at speeds over 40 m.p.h. the aerodynamic properties of the vehicle have a much greater effect. At 55 m.p.h. drag absorbs about 40 per cent of engine power. In Figure 91 you can see that at 70 m.p.h. the engine power consumed by air drag is greater than all other resistances put together.

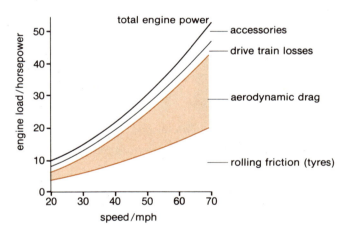

Figure 91 Effect of speed on power requirements for standard-sized car

However, even at lower speeds the drag shape is of some importance. The Volkswagen company has calculated that the proportion of air resistance to all other resistances on a Golf-class vehicle is as follows:

for town driving	18 per cent,
at 90 km/h	70 per cent,
at 120 km/h	80 per cent.

A Volkswagen prototype research vehicle aims to reduce the drag by 30 per cent, which would achieve the following *savings* in fuel:

for town driving	5 per cent,
at 90 km/h	16 per cent,
at 120 km/h	20 per cent.

Volkswagen calculated that over a distance of 100 000 km (including 50 per cent town driving) the fuel saving would be 900 litres.

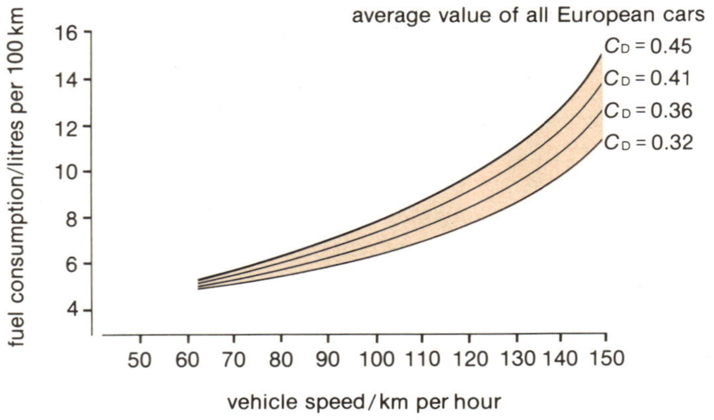

Figure 92 Comparison of fuel consumptions of four cars with drag coefficients of 0.45, 0.41, 0.36 and 0.32

Even relatively modest improvements in drag coefficient can give valuable savings. If the drag coefficient drops from 0.46 to 0.39, then a saving of 11 per cent is achieved at a cruising speed of 120 km/h (see Figure 92).

A rough guide to the fuel savings low drag can bring is to divide the reduction in drag coefficient by between three and five. Thus a 10 per cent reduction in drag gives possible fuel savings of 2 or 3 per cent. By this token, if it were possible to halve the drag coefficient, then a benefit of around 15 per cent in total fuel saving could result.

One designer, Luigi Colani, has re-styled the Citroën 2CV to give it a drag coefficient of 0.27 and claims to have achieved 100 m.p.g. (Figure 93). Is this likely by shape improvements alone?

Figure 93 Re-styled 2CV claimed to do 100 m.p.g.

The normal fuel consumption of the 2CV is 47 m.p.g. for touring. The published drag coefficient is 0.50. The Colani body represents an improvement of 0.24, say an improvement of 50 per cent, which can only give 17 per cent maximum improvement in fuel consumption, say 8 m.p.g., bringing the new consumption to 55 m.p.g.

Colani must use other means to reach a figure of 100 m.p.g.

It is important to keep the claims of aerodynamic designers in perspective, both those of individual inventors and of large companies promoting new models. Nevertheless, it is quite certain that in the next two decades aerodynamics will be a crucial contributor to both fuel saving and the outward appearance of vehicles.

Better late than never

In the television programme you will see some of the work of Jaray and his followers illustrated. Also, Geoffrey Carr of the Motor Industries Research Association is shown conducting wind-tunnel tests on a demountable model very much in the manner of Lay. Part of Geoffrey Carr's research in 1963 showed that by careful attention to detail, cars can have low wind resistance without looking at all streamlined. He demonstrated that the drag coefficient of a Renault Dauphine could be reduced to 0.25 by a series of simple modifications. Yet the work of the early theorists and these later more comprehensive demonstrations has largely been neglected.

Why have the principles of aerodynamics taken so long to sink in?

I think there are been three major inhibitions.
1 The uncertainty about the results of aerodynamic research.
2 The seemingly limitless supply of cheap fuel.
3 The market resistance to novel shapes.

Let's explore these factors in a little more detail.

Firstly, there is an internal resistance to the results of wind-tunnel testing. Designers have been able to claim that model tests do not always give a very accurate prediction of the behaviour of a full-scale production car. For example, the 1956 Porsche Spyder showed a drag coefficient of 0.29 in model form, but a later full-scale tunnel test registered 0.45. There are genuine worries about some wind tunnels, where the tunnel throat is small or the tunnel walls distort the airflow. Earlier I noted some of the approximations that are entailed in model testing. All in all, you could say that there is a little uncertainty in the results of aerodynamics research, and designers have found it convenient to ignore these tests. They are seen as a further unwelcome constraint. The debate can be interpreted as engineers and their rational principles encroaching on the freedom of manoeuvre of body stylists.

Tony Curtis, writing in *New Scientist* put it this way,

> This revolution in design is leading to some angry claims and counter-claims by conventional designers who reiterate the old excuses. One of these is that little value can be attached to drag values because, they say, 'it all depends upon what you mean by frontal area' – though it is difficult to imagine a less ambiguous concept. Another is that 'each wind tunnel gives a different answer', though a correlation exercise conducted a few years ago showed that the results obtained from the majority of the important European wind tunnels agreed closely.

> The battle of the drag factors is warming up, and the ordinary motorist will be the beneficiary – albeit some 20 years later than he should have been.

> (Curtis, 1981, p.549.)

Secondly, until the fuel crisis of the 1970s there was very little *stimulus* to aerodynamic design. For the period between 1925 and 1965 aerodynamics was a specialist interest. Important for speed, for record-breaking cars, perhaps, but otherwise of marginal importance to mass-produced cars, with one or two exceptions. For instance, the Chrysler Airflow of 1934, which did not match the promise of the tunnel tests, became a commercial flop. It was phased out in 1937 for both 'aesthetic reasons' and technical ones.

Figure 94 General Motors' recently constructed wind tunnel

The 43 ft *fan and* 5000 h.p. *motor can create* 150 m.p.h. *winds for testing vehicle aerodynamics*

Also, we saw that Jaray had an indirect influence on the Volkswagen 'Beetle', which is aerodynamic in its style (if not in its performance, with a drag coefficient of 0.485).

Only recently have the major manufacturers taken the subject seriously, constructing, for instance, their own in-house tunnel facilities (Figure 94). The impetus behind such expenditure, of course, is that great fuel savings could be offered. As explained earlier, a reduction in drag of 10 per cent, a relatively easy target starting from the current average 0.45, will give a fuel saving of 3 per cent. This figure includes city driving in a typical cycle. So the general savings to the consumer and the economy are fairly significant, especially when coupled with other fuel-saving technologies.

Thirdly, the general view in the car industry of aerodynamic shapes has been that such shapes will *not* sell. The designers embody the alleged views of consumers in the following kind of remark, 'If the possibilities were taken to the limit in engine efficiency, weight and aerodynamics, then the resulting car would probably look like a banana . . . and drive the same way'. (Designer quoted in 'Miles Ahead', Design Centre Exhibition, 1980.)

Let Tony Curtis sum up the position:

> Perhaps it is partly because low wind-resistance is popularly associated with such long and rounded shapes that the application of aerodynamics to motorcars has been so shamefully neglected. But fashion-conscious stylists and marketing men also have a lot to answer for, because well-informed engineers and motoring enthusiasts have known for decades that a great deal can be done to reduce the aerodynamic drag of the ordinary family car. The frontal area of a car cannot be reduced very much without forcing the passengers to assume uncomfortable postures, but designers can achieve respectably low coefficients without resorting to extreme or impractical shapes.
> (Curtis, 1981, p. 548)

The recent history of car design has shown more and more of the influence of engineers, and less of the old school of stylists. Apart from the fuel crisis, another external factor may have precipitated this change. The decline of the aircraft industry, particularly in the United Kingdom, has meant that many engineers migrated from the declining industry to the car industry. They brought with them their slightly different skills and preoccupations. It can hardly be an accident that the prime factors of aircraft design, weight and aerodynamics, now preoccupy the car industry.

SAQ 22

What is it that makes the ideal aerodynamic shape rather impractical for a mass-produced car?

6.5 Television programme 'The shape of cars to come'

The television programme deals not only with the future perspective for aerodynamic design but with some broader implications. The programme touches upon the relation between engineering requirements and marketing demands, the tension if you like, between the rational and not-quite-so-rational demands. It also examines briefly the history of aerodynamic research, and shows that the general principles of aerodynamics have been well understood since the experiments of Paul Jaray and others in the mid 1920s. It is puzzling that these principles have not been put into practice until recently, at least not in mass-production cars. In the programme the inhibitions on aerodynamic design are discussed.

The participants in the programme include:

Uwe Bahnsen, Vice President of Design in Ford, Europe, who shows how the accommodation of passengers and the safety requirements for accessories can militate against aerodynamics.

Ernie Thompson, Marketing Manager of Ford UK, who discusses the problems of introducing novel shapes into the market, and how the company needs to 'educate' consumers to accept novelty.

Sally Ford-Hutchinson, a market-research consultant to the motor car industry, who addresses further the difficulties of assessing public demands. The more general problems of what people want and what they will buy begin to emerge.

In the view of Tony Curtis, editor of *Motor* magazine, the tradition of the car industry is one of 'fashion', and only now are sensible engineering principles beginning to determine body shape.

Yet despite the proven advantages of aerodynamics, it tends to become another selling point, something to dramatize in the interests of advertising and marketing. The atmosphere surrounding the launch of a new car is well captured in the programme. The mixture of excitement, optimism, apprehension and religious tub-thumping may seem melodramatic to you. But the cost of designing and developing a new car is so large (£700 million for the Sierra) that no one close to the project can afford to think about its failure. Everyone is an evangelist for the new car.

Moreover, Ford, as a methodical and well organized company, do not stop their finance and their effort at the point when the car is complete. After it is researched, designed and developed, the effort continues in promotion and marketing. It would be foolish to spoil the ship for a £2.5 million-worth of tar. But when you are watching the programme, remember to keep a sensible scepticism about some of the claims and remember that what you are watching is the culmination of a five-year programme and a particular company's style.

Programme notes

Before the programme

You should read through this section 6. You should understand the main factors that contribute to aerodynamic drag, and some of the problems of modelling aerodynamics for cars. Also from this section you should have a broader understanding of the history of aerodynamics and of the benefits of aerodynamic shapes and their limitations. You should also have a good idea of the range of drag coefficients of current cars and the probable improvements to be achieved in the immediate future.

During the programme

Make a note of the different opinions given about aerodynamic design.

Try to distinguish between requirements that come from within, that is, from the designers, engineers and the company itself, and those requirements that come from outside, for instance from legislation or market research.

After the programme

Consider these issues:

Aerodynamic shapes. Think about the forms developed by Jaray and other historical examples. Try to come to a basic familiarity with aerodynamic shapes. Think about what factors prevented such shapes becoming widely used.

Technical factors. Think about the design of the Ford Sierra and the development of the prototypical Probe III. Uwe Bahnsen in the programme says that 80 per cent of the technical features are feasible now, so what is the main inhibition on taking them up? What example of safety legislation interacting with aerodynamics is used in the programme?

Marketing. The car industry has been described as part of the fashion business. Does the salesmanship associated with the Ford Sierra bear this out?

After you have reflected a little on the television programme try to answer the following self-assessment questions.

SAQ 23

In the television programme, Tony Curtis says the benefits of aerodynamics can be achieved 'without penalty'. What do you think this means?

SAQ 24

Give an example of the conflict between safety requirements and aerodynamics.

SAQ 25

What are the main dilemmas of marketing a new car like the Sierra?

PART THREE

7

FORECASTING THE FUTURE

We now come to the final part of this block, leading in to your assignment. We have looked at the past and the present of the motor car; we now turn to look at its future.

The final sections of the block will be concerned with some significant developments that are expected to influence the design and use of cars in the future, such as the search for alternative fuels as world oil supplies are gradually depleted. But first, in this section, I want to consider in more general terms how one can attempt to forecast the future and to use those forecasts in design.

Since designers are concerned with making proposals for things that do not yet exist, they are inevitably concerned with the future. In car design particularly, where there is a long time-span between the first ideas or proposals and the realization of those ideas in the manufactured product, there is a definite need to keep in touch with trends, developments and forecasts of the future.

In considering future forecasts in the context of cars, your first thoughts might turn to the sketches of the 'dream car of the future' that car designers sometimes produce. These conjectures are often simply visual hypotheses, with little technical reality. Nevertheless they can be quite influential in setting or expressing certain styling trends (Figure 95).

Another kind of conjectured 'car of the future' you might think of is the type that suggests some radical innovation such as electric cars, urban micro-cars, driverless computer-controlled cars, or cars designed for use in conjunction with rapid-transit systems. These visualizations (and sometimes prototypes) are often made in order to suggest how car design might change to meet changes in circumstances such as new laws or new technological possibilities (Figure 96).

Thus we can identify two main types of 'futuristic' car designs. One type of future car is the dream of hypothetical conjecture; the other is the 'necessity' of changing circumstances. These two types relate to what are regarded as two main kinds of futures forecasts in general: *normative* forecasts and *predictive* forecasts.

7.1 Normative forecasts

Of course, it is not possible to forecast the future with any real certainty. The best that one can do is to make forecasts either of what one *expects* to happen or of what one *would like* to happen in the future. Saying what you expect to happen means making a prediction, whereas saying what you would like to happen means setting some goals or *norms*, hence this latter type of forecast is known as a normative forecast.

Perhaps the most extreme kind of normative forecast is the Utopian novel. These normally describe the author's vision of an ideal state that could come into existence at some time in the future. (Alternatively, as in the original story of *Utopia* by Sir Thomas More early in the sixteenth century, the ideal state already exists, but in some far-off land.) Here, for example, is an extract from a recent Utopian novel (set in 'Ecotopia', an autonomous region of the United States of America), in which the author describes this future state's transport system.

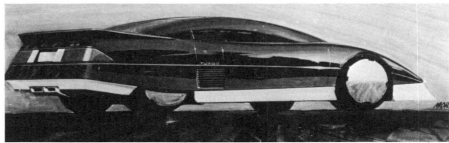

Figure 95 Dream cars
A selection of designers' fantasies

Figure 96 Car prototypes promoted
by the idea of electric traction

CAR-LESS LIVING IN ECOTOPIA'S NEW TOWNS

San Francisco, May 7 [1999]. Under the new regime, the established cities of Ecotopia have to some extent been broken up into neighborhoods or communities, but they are still considered to be somewhat outside the ideal long-term line of development of Ecotopian living patterns. I have just had the opportunity to visit one of the strange new minicities that are arising to carry out the more extreme urban vision of this decentralized society. Once a sleepy village, it is called Alviso, and is located on the southern shores of San Francisco Bay. You get there on the interurban train, which drops you off in the basement of a large complex of buildings. The main structure, it turns out, is not the city hall or courthouse, but a factory. It produces the electric traction units – they hardly qualify as cars or trucks in our terms – that are used for transporting people and goods in Ecotopian cities and for general transportation in the countryside. (Individually owned vehicles were prohibited in 'car-free' zones soon after Independence. These zones at first covered only downtown areas where pollution and congestion were most severe. As minibus service was extended, these zones expanded, and now cover all densely settled city areas.)

[. . .]

We toured the factory, which is a confusing place. Like other Ecotopian workplaces, I am told, it is not organized on the assembly-line principles generally thought essential to really efficient mass production. Certain aspects are automated: the production of the electric motors, suspension frames, and other major elements. However, the assembly of these items is done by groups of workers who actually fasten the parts together one by one, taking them from supply bins kept full by the automated machines. The plant is quiet and pleasant compared to the crashing racket of a Detroit plant, and the workers do not seem to be under Detroit's high output pressures. Of course the extreme simplification of Ecotopian vehicles must make the manufacturing process much easier to plan and manage – indeed there seems little reason why it could not be automated entirely.

Also, I discovered, much of the factory's output does not consist of finished vehicles at all. Following the mania for 'doing it yourself' which is such a basic part of Ecotopian life, this plant chiefly turns out 'front ends', 'rear ends', and battery units. Individuals and organizations then connect these to bodies of their own design. Many of them are weird enough to make San Francisco minibuses look quite ordinary. I have seen, for instance, a truck built of driftwood, almost every square foot of it decorated with abalone shells – it belonged to a fishery commune along the coast.

The 'front end' consists of two wheels, each driven by an electric motor and supplied with a brake. A frame attaches them to a steering and suspension unit, together with a simple steering wheel, accelerator, brake, instrument panel, and a pair of headlights. The motor drives are capable of no more than 30 miles per hour (on the level!) so their engineering requirements must be modest – though my guides told me the suspension is innovative, using a clever hydraulic load-leveling device which in addition needs very little metal. The 'rear end' is even simpler, since it doesn't have to steer. The battery units, which seem to be smaller and lighter than even our best Japanese imports, are designed for use in vehicles of various configurations. Each comes with a long reel-in extension cord to plug into recharging outlets.

The factory does produce several types of standard bodies, to which the propulsion units can be attached with only four bolts at each end. (They are always removed for repair.) The smallest and commonest body is a shrunken version of our pick-up truck. It has a tiny cab that seats only two people, and a low, square, open box in back. The rear of the cab can be swung upward to make a roof, and sometimes canvas sides are rigged to close in the box entirely.

A taxi-type body is still manufactured in small numbers. Many of these were used in the cities after Independence as a stop-gap measure while minibus and transit systems were developing. These bodies are molded from heavy plastic in one huge mold.

These primitive and underpowered vehicles obviously cannot satisfy the urge for speed and freedom which has been so well met by the American auto industry and our aggressive highway program. My guides and I got into a hot debate on this question, in which I must admit they proved uncomfortably knowledgeable about the conditions that sometimes prevail on our urban throughways – where movement at *any* speed can become impossible. When I asked, however, why Ecotopia did not build speedy cars for its thousands of miles of rural highways – which are now totally uncongested even if their rights of way have been partly taken over for trains – they were left speechless. I attempted to sow a few seeds of doubt in their minds: no one can be utterly insensitive to the pleasures of the open road, I told them, and I related how it feels to roll along in one of our powerful, comfortable cars, a girl's hair blowing in the wind. . .

(Callenbach, 1975, pp. 24, 25–6.)

In general, Utopian views of the future are short on technical descriptions (e.g. of the new machines or processes) and long on social prescriptions. Their authors are concerned with setting social goals and suggesting how people should behave and how society could be better organized.

Such Utopian visions often seem highly desirable, but not very probable. Frequently, any technical descriptions lack feasibility. After all, their authors are usually impractical novelists rather than realistic technologists. But technologists need to make conjectures, too, to offer alternatives and to suggest possibilities. 'Progress', as Oscar Wilde said, 'is the realization of Utopias.'

Here is a technologist, the engineer M. W. Thring, describing his idea for a 'multi-purpose trolley-seat', which could be the basic transport unit in a future society:

> Everyone could own a trolley, which would have on it a seat and a luggage box. There would be room for more luggage under the seat and the luggage box could be positioned either behind the seat to form a back-rest or in front of the seat to form a leg-rest, so that the seat could become a fully reclining couch. The trolley would have two small front wheels which could be locked to point forward or sideways. At the back it would have a single wheel on a castor so that, although the unit would normally be pushed forward, it could also be pushed sideways to manoeuvre it into a row of seats on a plane or 'train'.

> All trolleys would have safety harnesses and would be locked (usually facing backwards) to the floor of trains, planes or individual powered carriages by a high energy absorption plastically deforming steel helix, giving a maximum displacement of one metre. A worktable would be carried on the trolley and this could be fixed in front of the passenger at various heights by steel tubes on both sides of the seat.

> This basic trolley would be bought or hired by an individual owner and garaged at his home. The electrically driven carriages, however, would be community-owned, identical and interchangeable. Carriages would have both sides fully openable, the top section a transparent sheet

hinged at the roof and the bottom section a steel sheet hinged at floor level and forming, when it was opened outward and downward, a ramp up which the trolley-seat could be hauled by a cable driven by a motor on the carriage. Their storage batteries would be plugged in and recharged at cab-ranks and rail stations so as to be ready for the next user.

Powered carriages could be either individual, one-, two-, or four-trolley carriages for use as 'cabs' on the streets, or 'trains' boarded at 'stations' and running along a track or overhead monorail 'highway'.

(Thring, 1973, pp. 65–6.)

Utopian? Possibly. Feasible? Probably. As a professor of mechanical engineering, Thring presumably knows whether the technical details such as 'a high energy absorption plastically deforming steel helix' or the rechargeable 'storage batteries' are practicable in the near future. He has even gone so far as to design and make a mock-up of the trolley-seat (Figure 97). So the technological conjectures embodied in this particular normative forecast for the future of transport are probably soundly based.

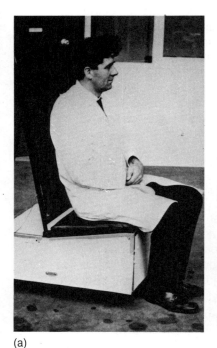

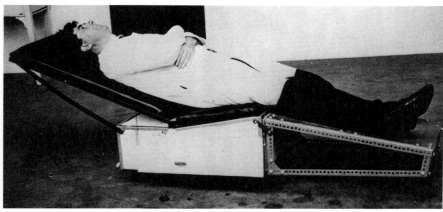

(a)   (b)

Figure 97 Mock-up of Thring's trolley-seat for a proposed international transport system

(a) The passenger would take the same trolley seat from starting point to destination, since it is designed to be pushed by hand or to be locked into position on individual power units or in buses, trains and aircraft
(b) The trolley-seat in reclining position to allow the passenger to sleep on long journeys

In his book *Man, Machines and Tomorrow*, from which the above extract is taken, Thring also discusses his ideas for the future of other major technological areas. He sets them all in the context of his vision of a future 'Creative Society', and therefore makes clear both his social goals and his personal values. Most other plans for the future of technology may be less radical, but are usually also less explicit in giving the values and goals of the planners. Yet any plan or design for the future is not simply saying what *will* happen (since the realization of the plan is not inevitable) but what *should* happen. It represents choices made between alternatives and therefore embodies value judgements.

The normative 'visions' of the future of people like Thring are useful because they offer alternatives to the prevailing assumptions about the future, and these alternatives throw a new light on current plans and forecasts that might otherwise be taken for granted. Even a very conservative plan can therefore be seen as merely accepting and adopting a particular set of values and goals. Radical alternative suggestions do have the merit of sometimes forcing the established planners to justify their own assumptions. For example, here is an extract from the report on *Traffic in Towns* prepared for the Government by the town planner Sir Colin Buchanan in the early 1960s:

It is possible, of course, if serious technological studies were undertaken, that a whole range of new ideas for moving people and goods in cities would be produced. It is indeed to be hoped that we are not at the end of our ingenuity in the matter. The bus, for example, for all its convenience, does not appear to be the last word in comfort. The travelator seems to offer much scope for development. Continuously operating chair-lifts might be used in a highly attractive way between points of pedestrian concentration to augment existing means of travel...

Even so it is difficult to see any new method of movement coming along which will be seriously competitive on a big scale with the motor vehicle. There are so many advantages in a fairly small, independent, self-powered and highly manoeuvrable means of getting about at ground level, for both people and goods, that it is unlikely we shall ever wish to abandon it. It may have a different source of motive power so that it is no longer strictly a motor vehicle, it may be quieter and without fumes, it may be styled in some quite different way, it may be guided and controlled in certain streets by electronic means, it may have the ability to perform sideways movement, but for practical purposes it will present most of the problems that are presented by the motor vehicle of today.

Our conclusion, therefore, is that the future of the motor vehicle, or of some equivalent machine, is assured.

(Buchanan, 1964)

Here, Buchanan rejects conjectures of radical innovations in transport technology, but accepts the social goals and personal values embodied in the 'advantages' of 'independence' etc. of private motor vehicles.

7.2 Predictive forecasts

You are perhaps thinking, 'But surely a report for the Government, prepared by a leading town planner, wouldn't, in any case, just make guesses about the future, or rely on Utopian conjectures? Surely it would use reliable forecasts, predictions of trends, and so on?' Yes, you are right. It relies in general on *predictive* forecasts, which try to outline what is expected to happen in the future if present trends continue. Plans are then made to cope with and allow for those expected outcomes.

For example, the Buchanan Report used graphs such as that shown in Figure 98, which forecast the expected rise in the number of vehicles in Britain. The graph was based on various assumptions about growth in incomes, motoring costs, patterns of vehicle use, and so on. Starting from the situation in 1963, it forecast that by 1980, for instance, the total number of vehicles in Britain would be some 27 million. In fact, there were 20 million vehicles in Britain in 1980. Obviously, some of the assumptions used in the equations in 1963 turned out to be wrong.

The idea behind this kind of trend extrapolation is quite simple. One looks at the past history of the trend, makes some assumptions about the factors that influence it, one prepares a mathematical equation incorporating those assumptions and uses the equation to generate a graph showing the trend's continuation into the future (or one simply extrapolates the trend graphically). The problems, of course, are that the assumptions might not be the right ones, the way they interact might not be the way the equation assumes they do, and changes in other factors in the future can have unpredictable effects. Figure 99 shows the projections of car ownership in Britain made by the Transport and Road Research Laboratory at various times. As you can see, the projections had considerable variations.

The lesson should be clear: even the apparent mathematical precision of trend extrapolation cannot show a *certain* future. Apart from any other factors, the extrapolations themselves can influence what actually happens after they are made, since the planners and decision makers can look at the graphs and decide that they have to do something to alter the trends.

103

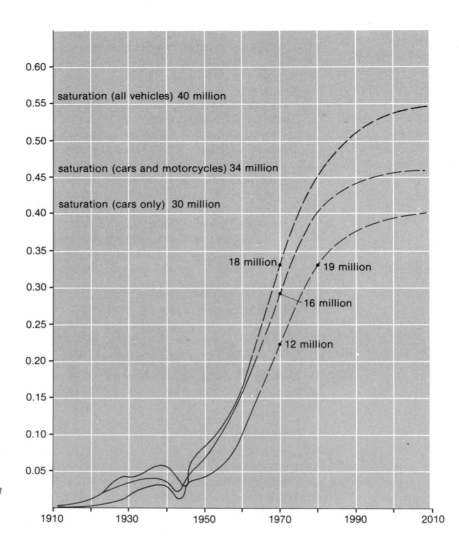

Figure 98 Graph of expected growth in number of vehicles, used in the Buchanan Report, 1964

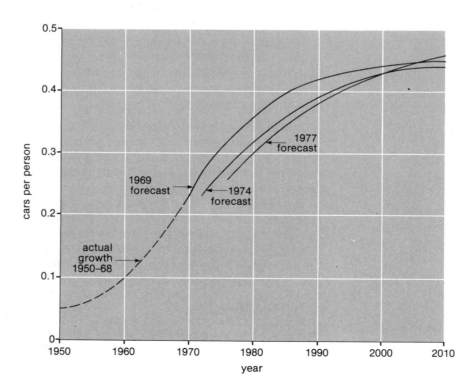

Figure 99 Various projections of car ownership in Britain made by the Transport and Road Research Laboratory

Because many factors often interact in a dynamic way to create unforeseen changes in trends, a more sophisticated kind of forecast uses a technique of 'cross-impact analysis' in order to consider together the future of related trends or events. If certain related events occur, then they may increase some trends and decrease others. In a full analysis, a cross-impact matrix is used to set out the potential interactions and to indicate the positive, negative or neutral effects that one change would have upon another. The probabilities of various events and changes are then computed. This can become quite complicated. A simple example might be an increase in fuel price leading to a decrease in car travel but an increase in motor-cycle travel, leading in turn to an increase in road deaths and injuries.

Another kind of predictive forecasting is to look at the way something is 'evolving' and to try to forecast the way it will presumably continue to 'evolve'. An example of this evolutionary forecasting is shown in Figure 100.

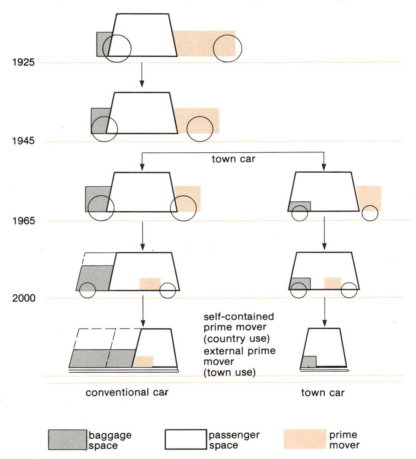

Figure 100 Evolutionary forecasting based upon a trend in the form of vehicles

7.3 Scenarios

A plausible 'scenario' for the future usually involves a mix of predictive and normative forecasts. 'Scenario writing', in this context of futures forecasting, means writing a plausible 'story' that starts at some specific date or event and proceeds to develop into the future in a realistic way using known facts and extrapolations, likely actions and reactions, explicit conjectures, and so on, until another specific date or event is reached. It therefore often combines a prediction of likely future events with an attempt to trace feasible paths to certain desired goals.

Here is a brief example of a scenario for developments in urban transport up to the end of the twentieth century, written in the late 1960s.

The last buses (electric, of course, because of the anti-pollution laws of about 1980) will have disappeared about 1990. Labour costs will have come to represent 90 per cent of the operating cost of buses, which will still be plagued by strikes of the dissatisfied drivers. Buses will finally become too heavy and cumbersome to be able to be integrated into the centrally and electronically guided traffic flow which will gradually extend to all the main arteries, and, by reducing the distance between electric cars to less than three metres, will increase the flow by a factor of 4.

Most ground level traffic will, however, be represented by small electric cars, cubic in shape. The urban branch of the car will have terminated its evolution and adopted the simplest most functional shape: that of a smooth plastic cube with no chromium plating and no proturbances, but standard rounded corners to facilitate handling and storing. The length of the cars will also be standardized to make them easily transportable in standardized containers. These silent, efficient cubes will have helped town-dwellers forget the noisy, evil-smelling vehicles that proliferate today. They will be propelled by high-speed electric motors running at 35 000 r.p.m. using alternating current without brushes or commutators. They will be fed at variable frequency by miniaturized thyristors. Transmission will still be necessary to convey the power to the microwheels. But such progress will have been made in construction materials that it will be possible to drive the wheels by a single-gear ratio with an efficiency of over 95 per cent. These successors to gears will be flexible, deformable and practically everlasting.

Even town cars will not for the most part belong to their users but will be rented. Self-service taxi cabs will be made available after legislation to reduce urban congestion.

The average town-dweller, when he leaves his office, will go to the nearest 'linear-mobile park'. The latter will be situated along the main arteries (ordinary parking will not be allowed). Self-service taxi cabs, the supply of which will be computer controlled, will be hooked magnetically on to a conveyor chain under the pavement, forming a slow procession as they await customers. The traveller will insert his credit card, made of magnetized plastic, into the appropriate slot and will be able to open the car door and sit at the steering wheel. Pressure on the accelerator will release his car from the conveyor.

If he is going to the suburbs, he will be able to drive himself, but if he is going to the university and has to cross the town via the main thoroughfare, he will have to integrate into the automatic driving system, after indicating his destination by word of mouth to the central computer . . .

(G. Bouladon, quoted in Waddington, 1977, p.205.)

SAQ 26

Try to identify the various features and techniques of both normative and predictive forecasts used in the above scenario.

The weakness of this 'scenario' is that it gives hardly any indication of how the desired changes are to be brought about. A scenario should include:

An indication of the time frame being considered (e.g. 'the next twenty-five years') and why (e.g. need to conserve resources within that period).

Identification of the factors, trends or events that are expected to be significant within that time frame for the particular area being considered. Obviously, it is impossible to include everything that *might* have some effect, and the scenario must therefore make some fundamental decisions on what to include and what to leave out.

A statement of the goals that are being aimed for in the scenario. Assuming that the scenario is essentially normative rather than predictive, the scenario writer should indicate what is the desirable state of affairs that he or she is trying to achieve.

An outline of the starting point for the scenario (e.g. the current state of affairs, the current trends and their extrapolations).

In this way, the scenario writer establishes the scenario starting point, its end point, and the factors or variables that are expected to influence the transformation from start to end. The scenario writer uses the different factor 'strands' to weave a plausible story. But it is not a 'story' in the sense of being a work of fiction; its plausibility lies in it having a sound basis of fact.

The following example of a scenario for the future of urban transport was originally written by J. C. Jones in 1959. Although written so long ago, the scenario still seems both plausible and relevant.

A plausible future for city transport

The road traffic example was written in 1959. It was then assumed that the crucial new components, i.e. an automatically controlled city car and a traffic guidance network, would not be feasible until 1980. Therefore the performance specifications were set to give a much better general performance than was feasible in 1959, e.g. vehicles to be available within *ten yards* of any point within the urban area at time delays of the order of *one minute* and *independent* of ice, snow and fog. An extract from the 1959 specification for an uncongested road traffic system appears below:

1. Facilities to Be Available to Road Users [...]
(a) Travel between points very close to origins and destinations
To move people or goods from points *very close* to their origins to points *very close* to their final destinations. This closeness is one of the chief advantages of road travel. To obtain maximum appreciable advantage it should be of the order of 10 yards. Origins and destinations can be at *any point* within the urban area.

(b) Very short delays at beginning and end of journey
To permit the beginning of a journey *very soon* after the decision to make it has been made and to permit the people or goods to reach their final destinations *very soon* after the journey is finished. This short time delay is another of the chief advantages of road travel. To obtain maximum appreciable advantage it should be of the order of one minute. Time delays are to be equally short at any time during the day or night.

(c) Services not affected by weather, visibility, etc.
The facilities noted here should be available regardless of weather conditions, visibility, and the like.

(d) Total journey times not exceeding durations of visits or times between deliveries
The total journey time should be of the same order as, or less than, the duration of the activity which the traveller has come to his destination to carry out. In the case of goods the total journey time should be of the same order as, or less than, the time between deliveries at the destination concerned. (It is suggested that these two criteria would be verified by an operational analysis of transport costs.)

(e) Total journey times not exceeding road users' expectations
The total journey time should never exceed travellers' expectations, i.e. it should always be possible to obtain an accurate estimate of arrival time.

107

2. Demands Made Upon Road Users [...]

(a) No bodily or mental stress remaining after a journey

No traveller should endure stresses that cannot be recovered from during the journey, i.e. there should be no residual stresses of mind or body. (It is suggested that this criterion would be verified by psychological and physiological studies of fatigue during road travel in relation to the expressed discomforts of drivers and passengers and to such long term effects as nervous breakdowns and stress disorders.)

(b) No learning required of road users

Actions expected of drivers and passengers should involve no new learning on their part and should therefore be compatible with the behaviour learnt in early life and should be well within the capabilities of the less capable members of the population.

(c) Low accident risk

There should be a very low probability of accidents to travellers or pedestrians.

3. The Cost of Road Using [...]

(a) High utilization of resources

Resources that are tied up in the system should have a high utilization so as to get maximum benefit from the capital cost of the system. Utilizations to be aimed at should be near to 100 per cent.

(b) Adaptable to changing conditions without obsolescence

The system should be adaptable to changes in traffic volume and vehicle performance without putting existing parts of the system out of date before they wear away.

(c) No rebuilding necessary

Improvements to the present system should not entail great changes to buildings and structures existing in 1959 or built since that date.

(d) No planning limitations

The system should not impose any limitations on the planning of places between which road travel is provided.

The new components that are compatible with this set of specifications are as follows:

1. All vehicles to be capable of both manual control and of automatic control by a cable under the road surface. Vehicles can move independently of road friction (so that they can travel in a straight line over ice or snow and can turn corners without slowing down).

2. All vehicles rented to users for duration of journey only.

3. Vehicles can be called to any kerb, and the journey time to get to any other kerb can be predicted. Traffic diverted to avoid congestion.

4. Traffic moves along existing streets in single streams at near constant speed and intersects without underpasses by slight adjustments to the speed of individual vehicles.

5. Little need for parking places as empty vehicles are automatically moved to places of high demand.

[...]

The system described above was chosen not only because it appeared to be physically and economically feasible by 1980 but also because there seemed to be at least one way of transforming the components of the 1959 system by stages to evolve it into the new system. Furthermore, each of the intermediate stages of evolution was chosen to provide substantial, rather than marginal, improvements in performance so as to benefit users from the start and also to provide an economic incentive throughout. Briefly, the evolutionary stages are as follows:

Stage 1 [...]
A largely improvised man-machine system for rapidly measuring and simulating the behaviour and cost of city traffic [...].

Stage 2 [...]
Immediate relief of congestion and easing of parking difficulties, by using the data handling system of stage 1 to control traffic lights and to broadcast information to vehicles trying to avoid congestion or to find parking places.

Stage 3 [...]
Existing vehicles phased out and new vehicles (compatible with both manual and automatic control) phased in.

Existing vehicles can hire radio links to receive individual routeing instructions and parking guidance from a traffic control centre. The phased-in vehicles provide an on-call hire service (dial-a-trip) from any parking meter or telephone and gradually replace private cars, buses and taxis. They are manually controlled in stage 3 but are compatible with automatic control in stage 4.

Stage 4 [...]
All vehicles are now of the new type and most roads are equipped for automatic control of fast single-lane traffic. Full automation permits continuous intersection of traffic streams at the same level and high utilization of vehicles that are automatically moved to kerbs where demand is greatest. The final system permits fast travel at little delay, with predictable journey times and in comparative safety. It is usable by a high proportion of the population, including children, old people and the physically and mentally handicapped.

Each state generates the know-how and hardware to begin the next and alters the economic balance to make it feasible to take the next step.

(Jones, 1980, pp. 320–23.)

7.4 Reading 12: Solomon Encel, Pauline Marstrand and William Page, 'The art of anticipation'

This Reading comprises extracts from the introductory chapters to the book *The Art of Anticipation*, edited by Encel, Marstrand and Page (1975). The editors are members of the Science Policy Research Unit at the University of Sussex. I should like you to study this Reading because it provides some background to the fairly recent development of 'technological forecasting' and contains some important comments about attitudes towards forecasting.

Now study Reading 12.

Why do the authors insist that forecasting is an art or a craft, rather than a science?

Discussion

The authors suggest that there are two principal kinds of writing about the future: that which is regarded as fiction and that which is regarded as forecasting. The latter stems from attempts to understand general social trends and the related planning for anticipated developments by Governments and other large organizations. In Western Europe there have been three basic kinds of planning for the future: town planning, economic planning and social planning. In Eastern Europe, planning has also been concerned with the restructuring of society, and the application of science and technology to specific social goals.

Forecasting, they suggest, has arisen as an attempt to make planning more rational and objective, hence more 'scientific'. However, they argue that forecasting cannot be a purely objective exercise: 'The future does not exist; forecasters try to invent it. In doing so, our presuppositions about what *ought* to happen are intertwined with assumptions about what *will* and what *can* happen.' There is no such thing as 'value-free' forecasting.

The authors are critical of forecasts which are guilty of naïve historicism – the assumption that the past controls the future – or determinism – the view that a single dominant force (e.g. technological change) determines the future. Instead, they want forecasting to be concerned with postulating and evaluating alternative futures, and with matching technological means to social ends: ' "Only in this way can we design technologies to serve human beings rather than extrapolate technical changes and think of ways to adapt human beings to these requirements".'

For these authors, then, forecasting is an art, or craft, in the sense that it involves subjective and imaginative ways of knowing and thinking. For myself, I would rather see forecasting as a design activity, being, in their words, 'an imaginative synthesis between what is known and what is unknown'. This synthesis is an essential element of the ways of knowing and thinking characteristic of design.

7.5 Reading 13: C.H. Waddington, 'Transport futures'

Finally for this section I recommend that you now read the chapter on 'Transport futures' reprinted from C.H. Waddington's book *The Man-made Future* (1978). It provides a broad context for your study of the remaining sections of this block.

Now study Reading 13.

Further reading

For keeping up to date with forecasting methods and applications, and a wide variety of other futures material, see the journal *Futures*, published by Butterworth Scientific Press.

Apart from the two books from which the two Readings for this section are taken, *The Art of Anticipation* (Encel, Marstrand and Page, 1975) and *The Man-made Future* (Waddington, 1978), other general introductory texts include:

E. Jantsch (1967), *Technological Forecasting in Perspective*, OECD.

G. Willis (1972), *Technological Forecasting*, Penguin.

Two popular overviews of the future are offered by:

A.C. Clarke (1974), *Profiles of the Future*, (2nd edn), Gollancz.

A. Toffler (1971), *Futureshock*, Pan.

A less frantic set of views is offered in:

C. Freeman and M.Jahoda (eds.) (1978), *World Futures*, Martin Robertson.

8
PERFORMANCE AND COSTS

8.1 Introduction

The motor car is an extraordinarily ingenious and complex artefact. It symbolizes some of the best capabilities of human beings: their inventiveness, their capacity to solve technical problems and their powers of collective organization. Yet the motor car also represents some of the worst characteristics of human beings: selfishness, inefficiency and their inability to control technological momentum, no matter where it takes them.

In the first two parts of this block the car was very largely accepted on its own terms, or rather on the terms of the car industry and those engineers and designers closely involved in the industry. But here we can step back a little and become more critical in our evaluation. To do this is not to demean the efforts and brilliance of many generations of automobile designers, but from our privileged external position we can see that one or two things are not quite right. If we could start again with Henry Ford in 1908, should we want the industry to come out the same way? Or are there more rational pathways of evolution? Firstly, the scale and effect of the industry seem overwhelming and irresistible.

It is hard to overestimate the importance of the motor car, not merely in its personal usefulness but in its place in the economy of the United Kingdom.

Here are a few statistics for the year 1980:

Total production of cars in the United Kingdom was just over one million vehicles.

Nine per cent of British exports were cars.

Six per cent of the British workforce were employed in the car industry, including 6000 engineers, 1570 scientists and about 6000 draughtsmen.

Consumption of steel by the industry ran at about 2500 tonnes per day (92.4×10^3 tonnes per annum).

Petrol consumed by cars was about 40 000 gallons per day (15×10^6 gallons per annum).

Similarly, it is hard to think of any area of life that has not been influenced or profoundly changed by the motor car:

There are about fifteen million cars and vans in use on the roads of the United Kingdom.

Fifty-eight per cent of British households have a car, with $1\frac{1}{2}$ million new car (and van) licence holders every year.

There are 339 483 km of roads in the United Kingdom and motorways take up 2452 km of that figure.

There are also the severely problematic effects of road vehicles:

Lead is released into urban atmospheres at the rate of 7000 tonnes per annum. Carbon monoxide is released at a rate one thousand times greater (8.3×10^6 tonnes per annum).

Finally, the motor car takes its dreadful toll in road casualties:

about 80 000 serious injuries per annum in the United Kingdom;

about 6000 deaths per annum in the United Kingdom.

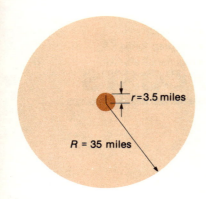

Figure 101 Area accessible in one hour's travel

Walking at 3.5 m.p.h., *accessible area* $\pi r^2 = \pi \times 3.5^2$ *square miles* = 38.5 *square miles*
Driving at 35 m.p.g., *accessible area* = $\pi R^2 = \pi \times 35^2$ *square miles* = 3850 *square miles*
Increase in accessibility = 100

Several important things emerge from this welter of statistics. The motor car is deeply embedded in the economy of the country. If the motor industry in the United Kingdom dwindles away then, not only is the effect on unemployment very severe, but the general balance of exports and imports is altered for the worse.

The hidden costs of the motor car in road construction, in urban areas, even in the layout and planning of ordinary suburban estates, are huge. One American writer estimated that the total cost is one-eighth of the gross national product in the United States. The penalties paid in pollutants and accidents for the benefits of the motor car are quite alarming.

So there must be very important benefits that make people put up with the ills of the motor car. One simple diagrammatic way to represent the benefits of the car is to think of it as a piece of equipment that gives its user access to a wider variety of possibilities. Look at Figure 101. This diagram assumes a uniform distribution of facilities (pubs, churches, friends, relations, shops or whatever). If you walk to these facilities, then in one hour's travel you have a territory of, say, 38.5 square miles from which to choose. But if you travel by car the accessible area increases to 3850 square miles. In other words, the choice you have increases a hundredfold. If you had access to one pub through walking, then you would have access to a hundred pubs by driving (assuming a uniform distribution of pubs). For that kind of reason it is hard to resist owning a car.

In this section I want to try to weigh up the benefits and deficits of the car in this kind of way. Also in this section I shall summarize some of the technical points of the block. It is intended to get you thinking about your assignment; at least to think about your point of departure. Pose these questions to yourself: Which areas covered in this block seem due for change? Which need changing?

8.2 Mechanical efficiency

The internal-combustion engine, you have seen in section 5, works at an efficiency of between 15 and 30 per cent. Even in the more efficient diesel engine only one-third of the fuel energy goes into usable power at the crankshaft. Another third goes into exhaust gases, and the remaining third into cooling losses. Some of the energy from the exhaust can be recycled by turbochargers, and perhaps some of the heat energy lost in the coolant can be retrieved, as in the adiabatic diesel engine (see section 5.3).

The lines of engine research we have covered in this block derive from:

increasing the ignition temperature,

raising the compression,

improving the burning of fuel and air,

recycling exhaust gases.

But we should think about the fuel efficiency of *the car as a whole*.

Look at Figure 102. On the left is the petrol in the tank giving the starting 100 per cent energy datum. Across the figure the progressive losses of energy are shown. For example, at the point of combustion 30 per cent of the energy goes directly into the exhaust gases, 30 per cent goes to cooling losses and only 40 per cent is available to propel the car. As you move downstream (think of this as an energy flow from the engine to the wheels) you can see how other losses occur.

I have labelled one large loss as 'transmission mismatch'. This is not a loss in quite the same sense as the others, as it refers to the inability of a conventional gearbox to take up the full power of the engine at all times. Thus

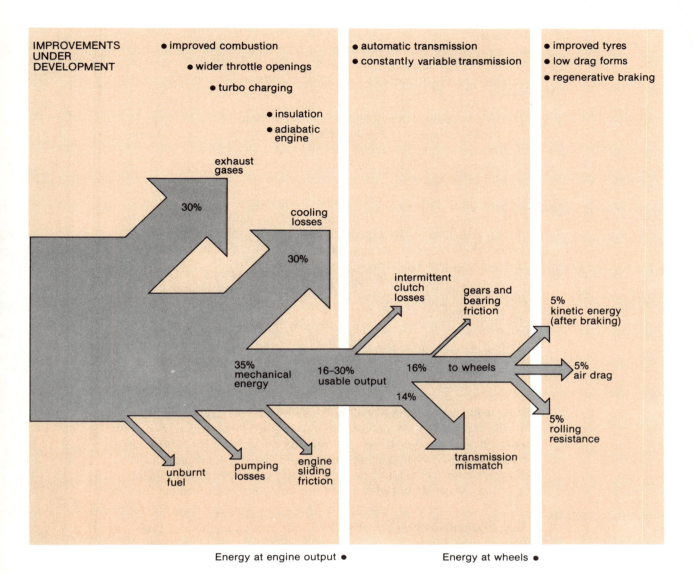

IMPROVEMENTS
UNDER
DEVELOPMENT

- improved combustion
- wider throttle openings
- turbo charging
- insulation
- adiabatic engine

- automatic transmission
- constantly variable transmission

- improved tyres
- low drag forms
- regenerative braking

exhaust gases

30%

cooling losses

30%

intermittent clutch losses

gears and bearing friction

5% kinetic energy (after braking)

35% mechanical energy

16–30% usable output

16%   to wheels

14%

5% air drag

unburnt fuel

pumping losses

engine sliding friction

transmission mismatch

5% rolling resistance

Energy at engine output ●   Energy at wheels ●

Figure 102 The dissipation of energy in a moving motor car

the mechanical output varies between 16 and 30 per cent, a variation of 14 per cent. A continuously variable transmission would remove most of that 14 per cent inefficiency.

Study Figure 102 carefully. It ties together many of the details described earlier in the block. Along the top I have indicated the most important potential improvements.

The kinetic energy, the energy required to push the mass of the car along, is a surprisingly small proportion of the total. Most of it is needed to accelerate the car up to its cruising speed. Once momentum has been built up, then the kinetic-energy demand is even smaller; for a car cruising at a constant speed in a straight line, it falls to zero. However, upon braking that energy is lost and once again the car must be pushed up to its cruising speed. You can see why the stop–go of urban driving is so wasteful of fuel energy.

It is important to realize that Figure 102 represents a momentary picture (a snapshot) of a car at a particular point in time. The distribution of the energy fluctuates all the time. So the widths of the various arrows would fluctuate.

For example, consider the first moments of ignition of the engine. Can you draw an equivalent map of the energy dissipation? Remember, the engine is turning, but the car is stationary.

113

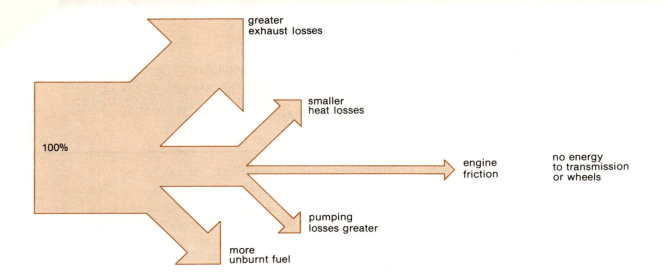

greater
exhaust losses

smaller
heat losses

100%

engine
friction

no energy
to transmission
or wheels

pumping
losses greater

more
unburnt fuel

Figure 103 The dissipation of energy in a stationary motor car, shortly after starting the engine

My version is shown in Figure 103. Because the engine is (relatively) cold there would be less heat loss, but a greater loss in exhaust and unburnt fuel: hence the smoke and fumes in the wake of a car that has just started. The engine friction would be about the same, perhaps slightly greater at half throttle (see Figure 62 in section 5). But more important, none of the energy reaches the wheels. It remains untapped, because the clutch is disengaged.

In the earlier parts of the block, I covered the major areas of potential improvement. All these technical improvements are the subject of current research and prototype development in the car industry. By 1990 most of them will have come into effect. You have read about some of these improvements in detail (engine improvements, transmission re-design and aerodynamic gains), but what does the total picture of the improved car look like?

Table 8 summarizes most of the information about potential improvements that are within reach.

Table 8 Summary of potential improvements over conventional late-1970s car

		Percentage fuel efficiency gain	subtotals
Drag	10% reduction in frontal area	2	
	20% reduction in C_d	4	
			6
Rolling resistance	10% weight reduction	10	
	change to radial tyres	3	
	or change to TRX type tyres	5	
			10–15
Engine	lean-burn engine	5–10	
	or stratified charge engine	10	
	or lightweight diesel engine	10	
	or adiabatic diesel engine	10–17	
			5–17
Transmission	automatic transmission with torque converter lock-up	10	
	or continuously variable transmission	10–15	
			10–15
Total possible improvements			34–53

Table 8 shows the balance and range of potential improvements in aerodynamic drag, in rolling resistance, in engine burning and transmission. These targets are *not* fantasy. They are quite realizable in the next decade; many have been achieved in prototypes.

Make a note of what strikes you as remarkable in Table 8.

You might have said that the biggest improvements are accessible through power-train design, in the heavier engineering, in newly devised engines and in transmission. However, the cost of such changes is much greater than others, such as improving the aerodynamic shape. This, therefore, is likely to be first in the line of the developments.

If I take a slightly longer-term view, an improvement of 50 per cent seems reasonably plausible. This strikes me as odd in the following way. By the mid 1970s, after a ninety-year history of invention and innovation, the last fifty years of which were very largely refining a well-established stereotype, the motor car was a very sophisticated, highly engineered artefact. Yet, by the measure of fuel efficiency, this artefact is only two-thirds as good as it should be. Under pressure from the fuel crisis, the designers, engineers and manufacturers, have discovered that they can do 50 per cent better than they have done in the past. What was once taken to be an efficient machine turns out to be quite inefficient.

This is not intended as a jibe at the millions of man hours of design effort that have gone into the motor car, but rather as an illustration of the imperfection and fallibility of the process. There is invariably room for improvement. This is a very important general lesson about design. The stimulus, as in the case of the car, usually come from an external source.

SAQ 27

Can you deduce the approximate speed of the motor car to which Figure 102 relates? It may help to refer back to Figure 91 in section 6.4.

SAQ 28

From the earlier account of various new designs and lines of research, would you say there is a close match between the work of the car industry and the greatest areas of fuel efficiency?

8.3 Performance gaps

As well as the mechanical inefficiencies of the combustion engine and the car as a whole, we should consider another kind of inefficiency, or at least a mismatch between the vehicle as designed and the typical performance demanded of it. The engine and car perform best at moderately high speeds, say 30–50 m.p.h. Yet, paradoxically, the bulk of car journeys (not car miles) are spent at low speeds in urban driving.

The average speed in London is 12 m.p.h. Not only that, but the percentage idling time, either stationary or at speeds below 4 m.p.h., in peak-hour driving is around 22 per cent. Clearly, there is a gross mismatch between the conditions of urban driving and the performance characteristics of a car.

115

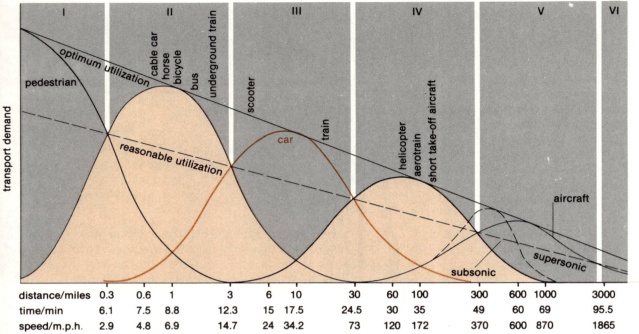

distance/miles	0.3	0.6	1	3	6	10	30	60	100	300	600	1000	3000
time/min	6.1	7.5	8.8	12.3	15	17.5	24.5	30	35	49	60	69	95.5
speed/m.p.h.	2.9	4.8	6.9	14.7	24	34.2	73	120	172	370	600	870	1865

Figure 104 Transport gaps

The gaps (shown shaded in brown) can be identified by plotting demand for transport, on the vertical axis, against the speed and optimum range of existing systems. The range is divided into six areas, of which I, III and V are well catered for by pedestrian, car and air transport. The notable gaps occur in areas II and IV. Demand for short journeys is much greater than for long (Source: Bouladon, 1967, p.42)

This mismatch is expressed diagrammatically in Figure 104. The transport demand is plotted vertically. There is a high incidence of pedestrian movements and a low incidence of supersonic aircraft journeys. Each form of transport has a typical performance curve. A car operates best in journeys of between 3 miles and 30 miles at average speeds of about 35 m.p.h. Of course, a car can operate at low speeds and shorter journeys; also it can operate at higher speeds. Its flexibility is a virtue and a vice.

On short journeys the car is performing inefficiently and on long journeys the combination of car and driver together performs badly through fatigue.

Gabriel Bouladon has identified what he calls 'transport gaps' on either side of the motor car. He sees a need for new forms of transport indicated by the brown areas on Figure 104. The first gap is for a vehicle that typically does journeys of a mile at a speed of around 7 m.p.h. The performance is much like that of a bicycle or horse. He also sees a need for a high-speed train, travelling at 150–200 m.p.h. to fill the gap above the motor car for long journeys. Since Bouladon wrote this article in 1967 high-speed trains such as British Rail's Intercity 125 have been introduced; moreover the use of bicycles in towns has increased significantly, as you saw in Block 2 *Bicycles*.

Bouladon also devised a system of transport for urban areas that would eliminate waiting time. This consisted of a continuously moving pavement travelling at 20 m.p.h. with an 'integrator' for access to the pavement and small compartment 'transporters' (Figure 105). Such a system is referred to in Reading 13 by Waddington.

Whatever you may think of the practicality and cost of such a system, it seems to me unarguable that there is something wrong with using cars for urban driving in the way they are now. The conditions created are both impractical and costly. A piece of equipment is designed and made with a 60 h.p. engine and a maximum speed of perhaps 100 m.p.h. or more, but is then used to travel a mile or so at 12 m.p.h.

SAQ 29

What kind of vehicle performance would most closely match Bouladon's transport gaps?
What kind of vehicles does that imply?

116

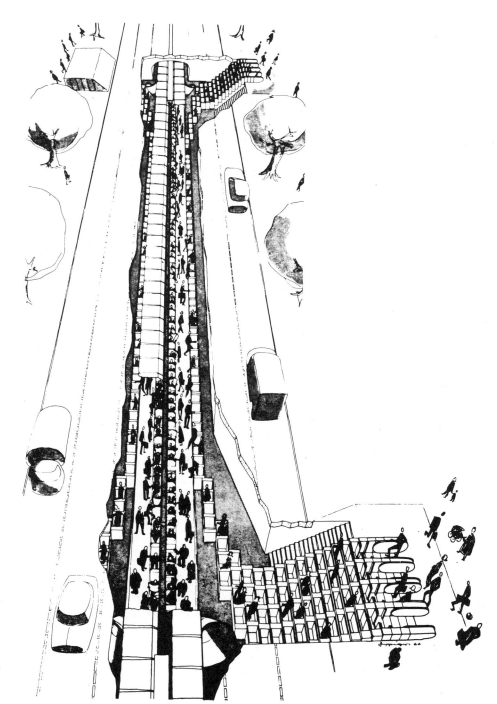

One alleviation of this problem is contained in the movement to smaller cars.
In America there has been a dramatic shift away from five- and six-passenger
cars.

Even this shift, however, leaves vehicle capacity and actual
transportation requirements badly mismatched. Surveys show that on
about 80 per cent of all trips American cars carry no more than two
people and that in a little more than half of all trips the driver is alone.
Therefore it is likely that if inexpensive, fuel-efficient two-passenger
cars become available in the 1980s, many will be sold. (Today the only
such vehicles available are relatively expensive and energy-inefficient
sports cars.)

It is hazardous, of course, to predict market behaviour when complex
social factors enter in, but assuming that periodic gasoline shortages
and price increases will continue to occur over the next decade, a

passenger-car sales mix in the mid to late 1980s might have a breakdown of something like the following: two-seaters 25 per cent, four-seaters 50 per cent and five- or six-seaters 25 per cent. In addition the demand for light trucks might drop to a ratio of only one truck to every six cars sold instead of the current one to every four. If the US automobile industry were prepared for such a shift, the industry might well benefit from it: some fuel-conscious households might choose to own an increased number of vehicles, each with a different functional design. A household that today owns two vehicles might decide, for example, to own three if they were more efficient than the present ones: two two-seaters for driving to work and for local errands and a third car with a capacity for four, five or six for family trips. Indeed, the rapid increase of three-vehicle households to about 20 per cent of all households in the 1970s is already due in part to the increasing popularity of small cars.

(Gray and von Hippel, 1981.)

So the American solution is to plug the lower transport gap with a range of cars, each household with two small cars for local trips and one big car for longer distances with the whole family. Is this realistic for the United Kingdom, where most people find just one car a costly business to sustain?

8.4 Financial costs

The cost breakdown of a typical mass-produced car is given in Figure 106. Figure 106(a) is a pie-chart of all the elements that contribute to the final selling price. The important divisions, shown in Figure 106(b), are as follows: dealer discount takes about 25 per cent, another 25 per cent goes into bought-in items (i.e. raw materials, sub-contracted parts and so on), 25 per cent is required for overheads, taxes and profits, while the remaining 25 per cent covers the fabrication and assembly of the car in the factory.

Moreover, the direct influence of the designer and the engineer is only in the last quarter. There is some indirect influence on bought-in items, but the remaining 50 per cent of the costs – profit, dealers mark-up, taxes and so on – is beyond the influence of the design team.

Manufacturing cost control linked to weight of components is one of the main methods that designers and engineers use to keep costs down. Yet their efforts are only a small part of the total picture. You may be able to think of ways to bring down the 40 per cent that is taken up by dealers, overheads and taxes.

The selling price of the vehicle, as far as the buyer is concerned, is only one part of the total cost of ownership. The cost of owning and running a car can be divided into three areas: standing costs, running costs and 'hidden' costs.

Standing costs are items like tax, insurance and optional subscriptions to organizations such as the Automobile Association.

Running costs include fuel bills, service and maintenance costs.

Hidden costs, naturally enough, are more difficult to gauge. The true cost of your car is only revealed when you sell it or trade it in, because only then do you know its depreciation in value. Also, the cost is related to how you pay for it. You may pay for a car outright in cash, or, more likely, you may take out a loan to be repaid in monthly instalments. The variety of purchasing patterns and the variety of ways cars are used makes it hard to draw general conclusions that would be true for most car owners.

Motoring Which?, April 1982, gives some typical cases of expenditure. These can be compared by taking the cost per mile of owning and using the car.

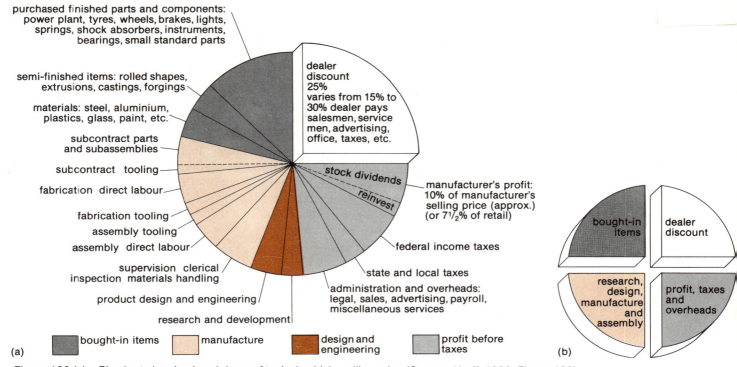

(a)

bought-in items   manufacture   design and engineering   profit before taxes

(b)

Figure 106 (a)   Pie chart showing breakdown of typical vehicle selling price (Source: Korff, 1980, Figure 133)
(b)   Selling price shown as four quadrants

Table 9 Typical examples of annual motoring costs

	15-year-old Mini; 3500 m.p.a.		2-year-old Renault 14TS; 9000 m.p.a.		17-year-old Triumph Vitesse; 2000 m.p.a.		Ford Escort Estate 27 000 m.p.a.	
vehicle licence duty	£77		£70		£70		£70	
insurance	£80		£138		£45		£140	
AA/RAC subscription	£30		£14.50				£30	
total standing costs	£187	5.3p/mile	£222.50	2.5p/mile	£115	5.75p/mile	£240	0.9p/mile
petrol	£165		£370		£115		£1350	
service & repair	£225		£50		£100		£150	
total running costs	£390	11.1p/mile	£420	4.7p/mile	£215	10.75p/mile	£1500	5.5p/mile
savings for replacement	£120							
bank loan repayments			£900				£1000	
depreciation					£40			
loss of interest	£40		£145		£65		£120	
total hidden costs	£160	4.6p/mile	£1045	11.6p/mile	£105	5.25p/mile	£1120	4.1p/mile
less company mileage costs			£100					
total motoring costs	£737	21p/mile	£1587.50	17.6p/mile	£435	21.75p/mile	£2860	10.5p/mile

Source: *Motoring Which?*, April 1982, p.221.

119

Two important points emerge for me from these typical cases. Firstly, the cost of motoring is more expensive than rail travel at 10p per mile for ordinary singles. Secondly, the cost of owning and maintaining a car approaches the cost per annum of buying house. For example, in 1982 a new Cortina at £5100 would entail a loan repayment of about £150 per month. But the value of the car depreciates, so either you lose money or you are compelled to trade in for a higher price the following year. Moreover, the original sum, if invested, would bring in £40 per month interest in a building society account.

No doubt you are aware that the cost of motoring is increasing. Not only are petrol prices rising, but the purchase price of new cars is also rising. Both are increasing *faster* than the rate of inflation. If you compare car prices between 1972 and 1982 you will see that they have roughly tripled in this period. During the same period the Retail Prices Index doubled, so of the 200 per cent increase, 100 per cent can be broadly accounted for as general inflation. The other 100 per cent *could be* attributed to improvements in design.

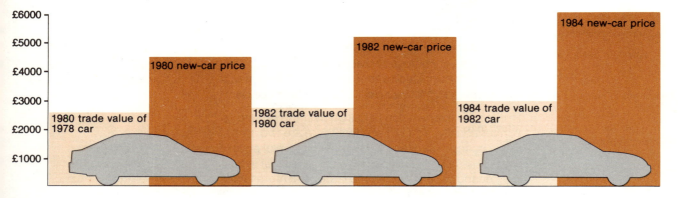

Figure 107 Increase in car prices, 1980–84

Some areas of improvement are clear, such as lengthening of service intervals, increased comfort, safety features, better performance, better fuel efficiency and better driving characteristics. But even with these improvements will consumers go on paying more? If consumers will not pay more, then product improvements can only be paid for by saving costs in manufacture and this can only be done by designing better processes as well as products. In a patents survey of Ford and BL for the past two decades, Ford was revealed as much better at developing new ideas and designs for both better products and better processes. An improved product subassembly design, which at the same time improves production efficiency, is likely to be one of the more important design strategies for the 1980s.

In total, the extra burden to the individual car buyer is the real increase in price, plus depreciation losses, plus loss of investment interest. This is not aimed to discourage you from buying a new car – perhaps you wouldn't dream of doing so anyway – but rather to put forward more general concerns. I should like you to think about the larger questions:

What does one get for the extra money one pays out for cars?

Is the car that costs a real 100 per cent more than it did ten years ago 100 per cent better?

Does it make sense to put money into a product that depreciates rapidly rather than into a construction that increases in value?

If consumers are willing to pay more, then what is the ultimate price ceiling? Repeatedly, in the post-war period, experts both in the industry and the press have intuitively felt that there is an ultimate price ceiling. Yet over the years product specifications and prices have continued upwards. Features of a car design that start as options after a few years become standard. However, this was during a general period of growth and expansion. If the recession continues and deepens, then stable price ceilings for each car class may emerge as a permanent feature of the 1980s.

8.5 Social costs

The social costs of any technology are notoriously difficult to assess, while benefits can usually be quantified into pounds and pence. For example, the argument for a new airport site would be accompanied by a schedule of the profits that would accrue to the airlines, the airport authorities and the local work force. But how would an estimate of the social costs of such a scheme be established? The hypothetical deficits may outweigh the hypothetical benefits.

The economist Edward Mishan suggested that for a true picture of costs and benefits to be realized, all the individuals affected by a new project (an airport, a motorway extension) should be asked, 'How much will you pay for this *not* to happen?' This could be conceived as an annual sum on the rates of householders. 'How much will you pay on your rates for this motorway on your doorstep *not* to be built?' By Mishan's argument the total of all these notional increases in rates would be placed against the projected benefits of the project.

But the social costs of the motor car resist such techniques, because its effects are very diffuse and sometimes not quantifiable. For example, how do you make a calculation about the effect of lead in urban atmospheres? Or how would you arrive at the social cost of a road accident fatality?

In this brief examination I shall go over the ground as dealt with conventionally. You will see that there are two areas that have generated major concern and legislation: environmental pollution and road accidents. In the first case one is assessing a nuisance that is costly in an intangible way. In the second case the costs are larger and more tangible. You will see that, indeed, financial estimates are made to try to come to terms with injuries and deaths.

Environmental pollution

Environmental pollution associated with motor traffic takes several different forms: noise, dirt, fumes and scrap debris. If we were to take a puritanical viewpoint, we might include the visual pollution of cars, particularly parked cars. Stationary cars in old towns and country estates form an unpleasant and (temporarily) useless kind of litter.

Conscientious attempts have been made to assess the nuisance factor of motor vehicles. A survey involving over 5600 interviews conducted in all kinds of residential areas in 1972 established that 64 per cent of the sample were bothered by motor traffic, a further 21 per cent were 'seriously disturbed' (Sands and Batty, 1974).

The nature of that disturbance varied according to the flow of vehicles and whether the respondents were at home or in the street. Figure 108 shows the type of disturbance that was found to be disturbing, plotted against vehicle flow. The general conclusions of the survey gave the following rankings for five disturbance factors:

1 pedestrian danger,

2 noise and vibration,

3 dust and dirt,

4 fumes,

5 parking.

Noise and vibration were attributed mainly (by 39 per cent of respondents) to lorries and only slightly (by 4 per cent) to cars. Fumes were not noticed much when indoors (by 7 per cent), but 47 per cent of all respondents were bothered by traffic fumes when walking in their neighbourhood, rising to 55 per cent in urban areas. Yet again, lorries were cited as the main offenders.

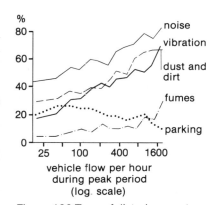

Figure 108 Type of disturbance at home related to traffic flow (Source: Sands and Batty, 1974, Figure II)

121

Clearly there is widespread concern about the bad environmental effects of traffic, but commercial vehicles rather than cars are perceived as the main culprits.

Fumes and noise are the subject of legislative controls. However, the nature and extent of the damage is uncertain. Environmentalists would argue that lead, unburnt hydrocarbons, carbon monoxide and oxides of nitrogen are highly toxic, and that the level of concentration of these substances in urban areas is unacceptable.

On the other hand, the motor industry sees that the tightening of emission controls, world wide from 1970, is both highly costly and likely to impede the development of fuel-efficient engines, such as the lightweight diesel (which generates high levels of oxides of nitrogen). Underlying that there is a resentment by engineers in the motor industry towards emission controls. This occurs because the recent history of legislation has, in their view, originated from the accidental circumstances of Los Angeles and its photochemical smogs. The strict controls of California in response to this special problem were rapidly followed by similar national and then international regulations. (See Figure 65 for the tightening of European controls.) From this point of view, one geographical anomaly does not merit this weight of restriction. Los Angeles' pollution was worse, perhaps by a factor of ten, than other major cities, such as Tokyo, Turin and Sydney.

Furthermore, the health hazard engendered by exhaust emissions is far less than other similar risks such as cigarette smoking. All of which is to say that there may be something disproportionate about the anxieties over exhaust fumes.

The problem is complicated by the fact that there are certainly pockets of urban environments that, at peak driving periods, suffer from very high levels of toxic substances. But general and national legislation is not perhaps the best way to attack local high levels of incidence.

The health hazards of car emissions then are topics of ambiguity and controversy. For example, in a 1978 study of Birmingham's 'Spaghetti Junction' the Department of the Environment asserted that there was 'no cause for special concern'. Yet one scientist involved in the study estimated that as many as 20 per cent of children under thirteen years in the city's inner area were affected by high levels of lead in the body.

Whatever the outcome of further research and analysis, the immediate future for the car world wide entails an increased tightening of emission controls, a related system of mandatory inspection and perhaps even specific local restrictions on automobile use.

Road accidents

Since 1945 there have been 209 244 people killed in road accidents in the United Kingdom. This number approaches that for the military fatalities of British and Commonwealth troops in the Second World War.

Happily, the recent annual totals show a decline. Despite the fact that $2\frac{1}{2}$ as many vehicle miles were travelled in 1980 as in 1960, there were 960 fewer deaths (6010 as opposed to 6970; see Figure 109). The sharp decline in 1973 and 1974 may be due indirectly to the fuel crisis, which led to less travel in vehicle miles (particularly by private motorists) and slower speeds by more fuel-conscious drivers. Statutory speed limits of 50 m.p.h. on ordinary roads and 70 m.p.h. on motorways may also have had an effect.

Table 10 Road casualties, 1980

Deaths	6010
Serious injuries	79 400
Slight injuries	243 200

The incidence of road deaths in 1980 (Table 10) was the lowest since 1958. However, the statistics still represent an appalling burden of pain and suffering. In particular, certain groups are highly vulnerable. For example, *half* the total deaths from all causes in males from fifteen to nineteen years are attributable to road accidents; while pedestrians over sixty years accounted

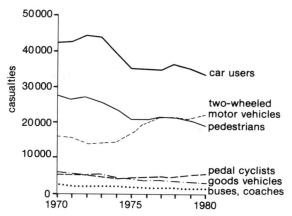

Figure 109 Fatal and serious casualties by class of road user, 1970–80 (Source: Road Accidents in Great Britain 1980, HMSO)

for 1105 deaths in 1980. Similarly, male pedal cyclists from ten to fourteen years account for 6069 serious injuries or deaths. (Incidentally this is over five times greater a casualty rate than for girls in the same age group.) Perhaps saddest of all, child pedestrians from five to nine years account for 11 122 serious injuries or deaths. The human cost is very great and incalculable. The economic cost is also great but, perhaps, can be calculated.

During the 1970s Government agencies, professional planners and researchers devised systems for calculating the cost of accidents. Their concern was to convert the damage arising from traffic accidents into a financial estimate. These calculations have two main purposes: firstly, to include 'accident costs' when evaluating, for example, alternative highway schemes; secondly, to make an estimate of the 'cost-effectiveness' of safety measures for preventing accidents.

So the techniques of costing traffic accidents can be viewed as a means of placing value on their prevention, and as an argument for investment in safety. This is the justification for the otherwise grisly exercise of putting a price on death. The costings that are used are roughly (1980 estimates):

fatal accident	£132 700
serious accident	£7080
slight accident	£970

The costs are made up by taking into account lost output of work, ambulance and hospital costs, damage to vehicles and property, police and insurance administration costs, and finally a notional sum to cover 'pain, grief and suffering'.

From these estimates the total economic burden to the United Kingdom of traffic accidents is around £2460 million per annum. The small reduction of 5 per cent in road deaths between 1979 and 1980 represents a saving of £40 million as well as a saving of 300 lives.

It is hard to establish the reasons for the decline in accidents while the number of vehicles in use continues to rise. Perhaps unpopular measures, such as alcohol laws, speed limits and seat belts, do have an effect. One survey in 1979 estimated that the efficiency of seat belts in averting death or serious injury is around 50 per cent for frontal and side impacts (Table K, para. 51, *Road Accidents Great Britain 1980*, HMSO). The legislation enforcing use of seat belts follows from such surveys.

Other benefits, no doubt, would follow from better designed highways and, indeed, better designed cars. Motor manufacturers are likely to be increasingly faced by the enforcement of tests on higher performance standards, and more defect recalls. In the United States of America, which has generally stricter legislation, during the period from 1966 to 1975, 43 per cent of automobiles (52 million vehicles) were recalled for safety defects. By such methods the economic burden of accident prevention is gradually being passed back to the motor industry.

SAQ 30

Use the 1980 estimate for the cost of fatal accidents to calculate the total cost of road deaths in the United Kingdom since 1945.

SAQ 31

The period from 1973 to 1975 saw a sharp decline in fatal and serious casualties to car users (see Figure 109). To what factor do you attribute that decline?

SAQ 32

Apart from physical danger, what other main hazards are generated by motor traffic?

9
ENERGY AND NEW FUELS

9.1 The energy context

Nothing disturbs the course of established design so much as a crisis or disaster. It stimulates and motivates a wider population than product designers alone to consider the possible consequences, and to demand or to suggest ways of averting them. Such an instance was the fuel crisis caused by the Fourth Arab–Israeli War of 1973 and its subsequent effect on world oil supplies.

For some years previously a few ecologically and conservationally minded thinkers had been warning of the eventual depletion of fossil fuels and many common resources. In 1968 a group of thirty industrialists, academics, national and international civil servants gathered in the Accademia dei Lincei in Rome to discuss the present and future predicament of man. They formed the Club of Rome, which initiated the study project on the Predicament of Mankind. In 1970 at meetings in Berne, Switzerland and Cambridge, Massachusetts they analysed the global patterns of the state of mankind on this planet. These were constructed into equations in systems models representing the population and economic growth, consumption and savings, and waste production. Professor Dennis Meadows at the Massachusetts Institute of Technology directed these computer modelling studies and the results were published in *The Limits to Growth* in 1972.

Also in 1972, the staff and associates of the *Ecologist* published similar findings, though with simpler models, in 'A Blueprint for Survival'. This document was endorsed by many eminent British scientists. There were many similar societies reporting their concern about resources and the energy crisis. Pollution itself was looming as a separate issue in the developed world.

All this stimulated thinking people to focus on four main issues: the growth in world population; the accelerating growth of production and consumption of the richer countries; the finiteness of resources and the rapid depletion of fossil fuels; increasing pollution and the destruction of environment by the direct results of industrial growth, its by-products and its wastes.

The direct concern for the dwindling stock of fossil fuels that could not be replaced was not at first in the forefront of the arguments. The optimistic expectations for North Sea oil produced figures of reserves, at times, a factor of five more than the eventual estimates. It took the oil crisis of 1973–4 to begin the change in people's aspirations and to prompt a re-direction in design trends.

To put the energy picture in perspective, let's briefly review the sources of supply and the destination of the fuels. Table 11 shows the contributions and splits for 1975, when the situation had settled down again.

It is clear from Table 11 that crude oil had become the major source of fuel, although only some 25 per cent of the 'oil barrel split' went to road transport (see Figure 110).

Table 11 United Kingdom energy flow, 1975

(a) Where it comes from

Source	10^9 kW h	
coal	858.5	36%
crude oil	1002.0	43%
natural gas	407.3	17%
nuclear	73.3	3%
hydro	14.6	1%
Total	2355.7	100%

Energy consumption of United Kingdom per head = 5.47 tce or 38 500 kW h.

(b) Where it goes

	10^9 kW h	
iron and steel	157.8	6.7%
industry	530.0	22.5%
domestic	433.4	18.4%
transport	355.7	15.1%
other	204.9	8.7%
waste	673.9	28.6%
Total	2355.7	100%

Note: This includes non-energy use of resources go ng to things like plastics, fertilizers, chemicals, animal foods, etc. Wastes are mainly power-station wastes and some refining losses.

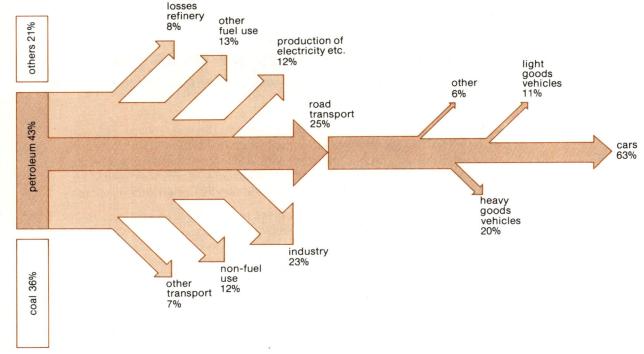

others 21%

petroleum 43%

coal 36%

losses refinery 8%

other fuel use 13%

production of electricity etc. 12%

road transport 25%

other 6%

light goods vehicles 11%

cars 63%

heavy goods vehicles 20%

industry 23%

non-fuel use 12%

other transport 7%

Figure 110 Energy used by transport in the United Kingdom, 1977

During 1973 the Arab countries were attempting to increase their participation in foreign oil companies in their lands, from 25 to 51 per cent. The Organization of Petroleum Exporting Countries was trying to increase its prices to protect against the rising inflation of Western countries. An increase from $3 to $4.20 per barrel was demanded. In November the Fourth Arab–Israeli War took seven days. Embargoes were placed on supplies to Israeli sympathizers, notably the United States of America and the Netherlands, which lasted into 1974.

This had a shocking effect on America and Europe. There were attempts to control consumption by fuel rationing, banning weekend driving of private vehicles, alternate days for supply at the petrol pumps, speed restrictions, banning of shop window lighting and illuminated advertisements. It led to brawls at the petrol stations and no supplies of heating fuels in some central states of America during that winter. But it did alert users and designers to the industrialized societies' addiction to oil and the possibility of some future difficulties in supply. It was the end of cheap fuel oils. The price was to continue to rise and was expected to do so throughout the 1970s until the world recession held the prices level at the end of 1981 (Table 12).

Table 12   Prices of crude oil and petrol

	Jan 1973	Nov 1973	1979	1980	1981	1982	1983
OPEC crude oil price/ ($ per barrel)	3	4–13	$13\frac{1}{2}$–$18\frac{1}{2}$	25–38	36–40	34	29
UK petrol price/ (p per gallon)		35	80–115	150–60	160–70	162–71	179

126

Remember that it takes from five to ten years to take a new car design from initial concept to first sales. How did the car designers of the mid 1970s react? There was some general feeling that things could not continue as before. They had to ask themselves, 'Is this just a hiccough or is the situation likely to get worse?' The dependence on oil supplies determines how long there will be a car industry, or does it? If there is to be increasing scarcity, competition will continue to push up prices and make transport expensive. These, then, were the general worries.

What now is the current estimate of oil reserves? The most reliable estimates of resources are published as the annual *World Petroleum Report*. Data are collected from all countries and producers. It is thought that all new oilfields, such as Alaska, the North Sea and the Mexican, create optimistic estimates, and all the older fields, such as American and Middle Eastern, are pessimistic.

Table 13 gives figures extracted from these reports. By dividing the current consumption into the proven and probable reserves we are able to obtain a rough figure for remaining supply years and the run-out date. There is a small error because the reserves are stated for the beginning of the year but the consumption is for the full year.

Some early figures come from other sources. In the early 1950s the world might have expected to run out of oil in about 1978, had there not been extensive exploration and finds in the 1960s. However, the same expectation does not apply now. The additions to reserves have been increasing at about $3\frac{1}{2}$ per cent per annum but consumption has been increasing at 7 per cent per annum. Some newly industrialized countries have growth rates far beyond this. Japan increased its oil intake by 17 per cent for four or five consecutive years. The other overriding fact is that we can now see 'the barrel emptying' by the fall from the peak reserve figures.

Table 13   World oil reserves and consumption

Date	Probable & proven reserves (A) /10^9 barrels	Annual world consumption (B) /10^9 barrels	Years to run-out A/B	Run-out date
1938	31 proven			
1945	58 proven			
1952	117.4	4.5	26.1	1978
1955	154.5			
1960		7.81		
1961	302.4	8.25	36.6	1997
1965	446.3	11.02	40.5	2006
1966	486.0	12.05	40.3	2006
1967	495.7	13.05	38.0	2005
1968	509.9	13.97	36.5	2005
1969	509.9	15.33	33.3	2002
1970	540.6	16.88	32.0	2002
1971	620.7	17.80	34.9	2006
1972	641.8	19.23	33.4	2005
1973	672.3	20.69	30.8	2005
1974	634.7	20.44	32.9	2007
1975	720.4	19.90	33.0	2008
1976	664.4	21.15	31.4	2007
1977	649.0			

Source: Mainly *World Petroleum Report*.
Conversion between reports at 7.35 barrels per tonne.

When car designers had eventually assimilated these trends, they had several clear routes for research and development:

to improve the performance of conventional cars through changes in their mechanics and their layouts;

to design a car that could use alternative fuels;

to design a different power source (prime mover).

In addition, the industry as a whole could adopt different strategies by:

improving energy efficiency in manufacture;

developing a long-life car;

improving recycling of car scrap.

I have covered the conventional solutions in some detail in Part Two of the block. In this part I shall discuss a couple of more speculative options.

9.2 Alternative fuels

New fuels

The fuels that might replace petrol can be thought about in two ways, as their original sources or as their delivered form. As far as the motor industry is concerned it would be convenient to have a delivered form of energy that is as close as possible in its characteristics to petrol. This would mean less change to the engine of the motor car. It also implies that the energy per unit weight (or energy density) of new fuels should ideally approach that of petrol.

The original sources of new fuels vary from sugar-producing crops, to wood, to organic waste, to sea water, to the traditional fossil fuels themselves. The variety of source materials and the variety of fuel forms may be a little confusing to you. You may come across names such as gasohol, ethanol, LPG, hydrogas and so on. Figure 111 tries to lay out the more important of these types.

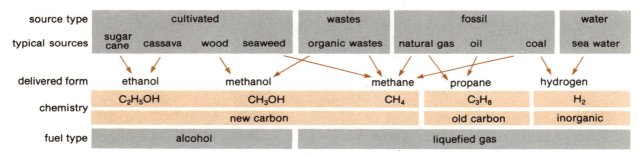

Figure 111 New fuels: sources and types

Essentially, there are two types of new fuel available to the motor car: alcohol and liquid gas. There are four types of fuel sources: cultivated crops, organic waste, fossil fuels and sea water. Sea water can be used to produce hydrogen, which is very different in its chemical properties from the carbon-based fuels. As a fuel source hydrogen is very speculative. It can be produced economically from sea water only if cheap electricity is available. Moreover, hydrogen is difficult to store either as a liquid or as a gas, but one line of research seeks to store it in a porous metal alloy, such as titanium/iron hydride.

On the organic front, lines of research include the cultivation of giant seaweed (or kelp). In the United State of America the General Electric Company has established an underwater farm off the Southern Californian coast. Kelp grows at a rate of up to two feet per day and it is hoped it will be digested readily to produce methane gas. A similar scheme by NASA has been running since 1975. In this case water hyacinths are used. This plant acts as a natural cleanser for domestic waste water, and it can be converted to methane and fertilizer.

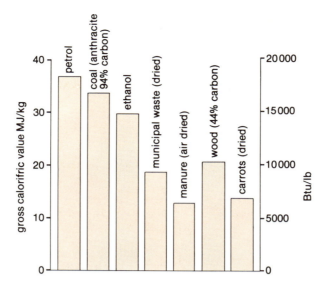

Figure 112 Gross calorific values of various fuels

However, in this section I shall look at those new fuels that are more developed, that contain 'new carbon' (see Figure 111) and which show some economic potential, as well as energy potential.

For comparison, look at the energy potential of some typical and not-so-typical fuels (Figure 112). Energy potential is expressed as gross calorific value, that is, the heat energy available from burning the basic material. Fossil fuels generally have higher calorific values than organic fuels, but, unlike fossil fuels, plants are renewable sources of energy. In theory, if properly managed, plant fuels should be inexhaustible. Some fraction of the plant material *should* be accessible and convertible into energy by an efficient and economic process.

What are the main candidates for plant fuels? What kind of efficiency can we expect in energy extraction from plants? What are the cost and trade implications?

There are several leading contenders for green power. The most developed of these are: sugar cane, cassava, maize, and, more speculatively, gopher trees (*Euphorbia*) and nut trees (*Pittosporum resiniforum*).

From the first group of plants the carbohydrates (sugar or starch) are converted to alcohol by fermentation and distillation. If you are a home brewer you will be familiar with parts of this process. Also you will have a sense that some crops are more suitable than others by virtue of their high alcoholic yield. But it is important to distinguish between two processes:

the fermentation of sugar by yeast, which produces *ethyl alcohol* (ethanol);

the destructive distillation of crops such as wood, or organic waste, which produces *methyl alcohol* (methanol).

These two processes eventually yield distilled and refined motor spirits. The characteristics of each fuel, ethanol and methanol, are given in Table 14.

There are inherent disadvantages in both these new motor fuels. As you can see from Table 14 the energy density of methanol is about half that of petrol, while ethanol is slightly better at two-thirds. Further disadvantages include low volatility, corrosion of fuel lines and miscibility with water that can generate vapour locks.

However, the high octane rating of these fuels gives the opportunity for a more efficient engine with higher compression ratios than usual. The attractiveness of alcohol for most countries is that they could conceivably generate their *own* fuel, instead of relying on oil imports. Let me summarize recent developments world wide.

Table 14 Characteristics of methanol and ethanol

	Methanol CH_3OH	Ethanol C_2H_5OH
Advantages	high octane number (high compression ratio)	high octane number
	low deposits	low deposits
	smooth operation in lean fuel mixtures	smooth operation in lean fuel mixtures
	high latent heat of vaporization (easier compression)	high latent heat of vaporization
Disadvantages	half energy density of petrol	$\frac{2}{3}$ energy density of petrol
	corrosion	corrosion
	difficult to start cold	difficult to start cold
	vapour locks	vapour locks
Raw materials	coal, wood,	grain, sugar cane,
	natural gas,	cassava, etc.
	organic waste	(ethylene as a petroleum by-product)

Source: *Ford Energy Report*, 1982, p.11.

Brazil leads the way in making alcohol from sugar cane. The Brazilian Government aims to reach a production of 10.7 billion litres per annum by 1985. In 1980 conversion of sugar cane produced 4 billion litres of alcohol. Many observers think that the 1985 target is not realizable. Nevertheless, by that time alcohol could be substituting for some 25–30 per cent of the country's petrol consumption. Most petrol pumps in Brazil already dispense a 20 per cent alcohol–petrol mixture. Higher ratios of alcohol require re-design of petrol pumps and fuel lines in cars because of corrosion problems. But even so by 1980 there were about 350 000 pure alcohol cars on Brazilian roads.

Similarly, Australia has considered developing its sugar cane crop for alcohol production. About 95 per cent of Australian sugar is produced in Queensland; by doubling the crop to about five million tonnes of raw sugar enough alcohol could be produced to provide 12 per cent of Australia's fuel requirements.

Very little alcohol is produced from cassava at present, although again Brazil leads the field as the largest producer of cassava in the world. Unlike sugar cane, cassava will grow almost anywhere in a warm climate, even on very poor soil. The Brazilian Government plans to plant cassava on two million square kilometres of poor agricultural land in the Gerrado region. However, the starch content of the crop has to be converted to sugar prior to its fermentation and distillation. This means some loss of efficiency.

In the United States of America the most obvious candidate for fuel is maize, or corn. A blend of 1:9 of corn ethanol to petroleum is sold already as 'gasohol'.

The net energy ratio (NER) of such a process is given by

$$NER = \frac{\text{final yield of energy}}{\text{net energy inputs}}$$

Therefore for a reasonably efficient process the NER should be greater than 1.0. It is estimated that the NER for corn ethanol is 1.55. The bulk of the energy input is in distillation and cooking. However, the energy that goes into cultivation and harvesting is not included. It is estimated that the NER would move below 1.0 if they were included. This means that the corn crop can only be converted into fuel by the use of less valued fuels such as local oil, natural gas or crop waste matter. As it now stands the energy equation does not make much sense: you put in more than you get out. It only makes sense if the process is viewed as the *conversion* of low-grade fuel into high-grade fuel. The use of corn as a fuel is perhaps more a symptom of short-term expediency than a rational energy policy, a response to crisis. This attitude is understandable on a local level, but it does not seem rational in broader terms—rather like burning the furniture to keep warm in winter.

A better mid-term option for the United States is wood (if not furniture), which already contributes 2 per cent to the nation's energy consumption. Of course, this is largely by direct combustion in domestic wood-burning stoves or larger chip-burning boilers. However, wood could provide a basis for making methanol. Experts calculate that it could cost about 65c a gallon to distil methanol from wood, whereas ethanol from corn would cost $1.25 a gallon. This may make methanol competitive with oil produced at 75c a gallon (late 1982 price). The biggest problem is the high cost of distillation works. The cost for such distilleries is between $500 and $1000 million. Few bankers are willing so far to invest in such a speculative venture. Another factor that is inhibiting is the high cost of distribution. Wood is expensive to transport, more so than coal. Moving wood more than fifty miles or so from its source makes it uneconomic.

Despite the handicaps and marginal gains, the major manufacturers have developed prototype cars which run on ethanol and methanol (Figure 113).

Alcohol cars

The results of prototype tests on cars using alcohol fuels are encouraging. They show a reduction in energy consumption and a reduction in exhaust emissions. For example, tests by Volkswagen on two engines running on pure methanol showed 20 per cent less energy consumption. Moreover, the high octane rating of methanol further improved the efficiency by raising the compression ratio from 8:1 to 13:1.

A New Zealand study of 1978 concluded that 10 per cent blends of methanol and petrol could be introduced for existing cars without engine modifications and without significant changes in fuel economy or driving patterns. Against their expectations there were no major problems presented by cold starting and vapour lock.

Figure 113 Methanol-fuelled Ford Escort

The engine of this car is extensively modified to overcome the deficiencies of methanol and to capitalize on its benefits. Extensive pre-heating of the intake is required to overcome poor volatility and the compression ratio is raised to 11.4 to 1

The Ford Company has vigorously pursued alternative fuels, especially in the Corcol as manufactured by Ford in Brazil. Their ethanol engine has a compression ratio of 13:1 with an exhaust-heated inlet manifold. Various electrical heaters have also been investigated to assist cold starting below 10 °C.

A fleet of forty methanol-driven North American Escorts has been developed by Ford. In these cars the compression ratio has been raised to 11.4 to 1. Methanol, like ethanol, needs about eight times as much heat as petrol to vaporize, so again water heat and exhaust heat are applied at the inlet manifold. Additives such as isopentane (5.5 per cent) give improved volatility. Corrosion is still a major difficulty and fuel lines of the forty vehicles will be made of stainless steel.

This research by Ford has shown that the methanol car overall is about 65 per cent as efficient as the petrol-driven equivalent, and the kind of modifications indicated here would cost the car buyer an extra $2000 per vehicle.

What are the reasons for optimism about alcohol fuels?

The loss of energy entailed in the change from petrol to alcohol fuels can be compensated for, in part, by improving the compression ratio. So on the swing of calorific value we may lose 50 per cent, but on the roundabout of engine efficiency we gain back 25 per cent. These engines also give much lower emissions of carbon monoxide, unburnt hydrocarbon and oxides of nitrogen. Moreover, the problems of cold starting, vapour lock and so on, do not seem to be insuperable.

However there do seem to be higher costs in solving the technical problems. The bulk of fuel required presents a major difficulty. For methanol a car would need a tank twice as big as normal. By comparison the weight of a titanium/iron hydride tank would be about five times as heavy as the equivalent petrol tank.

Implications

If there were a general movement towards these alcohol-based fuels, there would be political, economic and technological implications. I think there are three areas of concern:

land competition;

trade fluctuation;

efficiency of new fuels.

Let's look at these factors in a bit more detail.

Competition for land. The ideal requirements for a fuel plant are that it should be grown on land (or sea) otherwise not suitable for any other crop or foodstuff. This is, in practice, very difficult except in special cases like latex-yielding desert trees, and seaweeds. All countries have a finite amount of cultivatable land and climates that limit production of all but specific crops. What area of land would be required to cultivate fuel feedstocks? The Volkswagen research group has made the calculations shown in Table 15.

Table 15 Area demand for the operation of one alcohol car during one year

Assumption: mileage 15 000 kilometres/year; fuel consumption 10 litres gasoline equivalent/100 kilometres; energy consumption 20% less than a gasoline car.

Biomass	Fuel	Area/ha
sugar beet	ethanol	0.42
sugar cane	ethanol	0.47
cassava	ethanol	0.66
potatoes	ethanol	0.56
wood	ethanol	0.55
wood	methanol	0.34–0.46
algae	methanol	0.14
wood (hybrid poplar 30 t/ha year)	ethanol	0.22

Source: *Bernhardt* (n.d.)

If all the fifteen million cars in use in the United Kingdom were to switch to ethanol derived from crops, such as sugar beet and potatoes, then 75 000 km^2 of arable land would be needed. This figure is about one-third of the total land area of the United Kingdom and substantially more than the 69 910 km^2 already under cultivation for crops.

But even if this fuel production were to come about on a small scale, then food prices and fuel prices would become linked. For instance in the United States of America, 'using corn to make fuel on a large scale is a surefire prescription for inflation in food prices. A common rule of thumb in the nation's cornbelt is that each 1 per cent increase in corn consumption leads to a 2 per cent rise in its price' (Bylinsky, 1979, p. 79).

So, even if the economic problems of cropping and distillation were solved, the price of foods derived from corn would increase. A further anxiety is that the shift to alcohol crops would promote 'monocultures'. Already in Brazil small farms are being displaced by sugar cane plantations for the alcohol programme.

More than that, in a world where there are high levels of malnutrition, is the morality of growing edible crops to power motor cars suspect? For example, envisage an increase in world oil prices that makes corn-based ethanol plausible as an economic competitor. The American corn surplus then becomes consumed internally as fuel. The contruction of very costly distilleries make it unlikely that the crop can revert to being a food rather than a fuel. The process is one-directional. This is an argument outside whether or not America chooses to sell, or give, its surplus crops to other countries. In future it may not have the choice, if its car industry comes to depend on a corn-based fuel.

Trade fluctuations. The plausibility of alternative fuels is determined by the price of oil. In other words, there is an economic threshold set by OPEC, only above which do certain new technologies become attractive alternatives. Some new fuels work within a safe margin: in Brazil the pump price for sugar-cane-derived alcohol is about half that of alcohol–petrol mix (1:5). The economics work in favour of alcohol even allowing for its 25 per cent loss of efficiency. On the other hand, the estimates given for fuel prices from the gopher plant (*Euphorbia*) run between $25 and $250 per barrel, that is, between the plausibly economic and the wildly uneconomic.

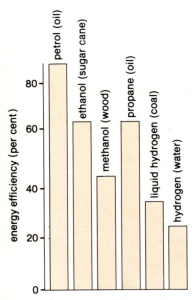

Figure 114 Energy efficiency of synthetic fuels

The wish to encourage alternative fuels can lead to strange anomalies. In 1979, the United States' national and state taxes usual for petrol were not enforced for ethanol, to the tune of $1 per gallon. Thus to avoid paying $21 for forty-two gallons of OPEC oil, the state lost $42 in taxes on forty-two gallons of ethanol.

The efficiency of the new fuels. The efficiency of these various new fuels has to be assessed against the high efficiency of petroleum spirit refined from oil. The cost of refining, the cost of tranport and the energy density of the fuel all have to be taken into account.

Look at Figure 114. Here we can see that ethanol, propane and hydrogen gas from coal are all on a par, but still at two-thirds the efficiency of petrol.

Also, hydrogen gas from coal might compete with other users, such as coal-burning power stations. So the kind of perspective that could emerge is a move of power stations to other forms of energy (hydro, solar, wind or nuclear) in order to release both oil and coal reserves to power motor cars.

Ethanol and methanol look more satisfactory in the sense that they derive from renewable resources. But ethanol may take up land that could be used for food production.

SAQ 33

From the information in this section how would you rank the state of development of the various new fuels?

SAQ 34

What are the factors to consider in the potential of a new fuel of the kind described here?

10
ELECTRIC VEHICLES

Whatever the pattern of the future, we can deduce that the main thrust of technical development is towards certain unchanged general aims:

to preserve the present degree of mobility afforded by the motor car;
to keep fuel prices relatively low;
to sustain the car industry as currently constituted.

The industry may or may not be successful in meeting these aims, but the general predominating trends are to make cars much as they have been, only more efficiently. In other words, most innovations will be gradual and incremental. This does not prevent us from thinking about more radical changes, such as alternative fuels.

Such speculations will, inevitably, be uncertain, not just because they presuppose more radical shifts in technology, but because their probability depends less on designers and engineers at the centre of the industry and more on politicians, economists and scientists. For instance, some future scenarios for the car industry depend on political decisions about nuclear power. Others depend upon chemists now working on new fuels or in battery technology.

How can the main choices be represented? There seem to be three basic options:

The present car can be substantially improved to reduce fuel consumption, and so make the remaining oil reserves last longer.

Alternative economic substitutes for oil and its derivatives can be found. Some of these may be from conventional sources, others from the less conventional sources described earlier.

Other forms of traction could be devised, which would rely on a different prime mover, and so replace the internal-combustion engine and its dependence on liquid fuels.

Arranged in this way the options step further and further from the car as we know it. They can be thought of as short-term, mid-term and long-term targets. (See Figure 115.)

In earlier sections of the block we have looked at the first option in some detail. The last section dealt with some of the possibilities of the second option. Let's now look at the third option.

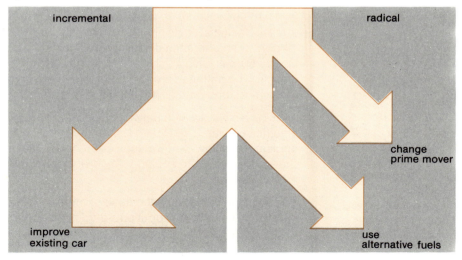

Figure 115 Main pathways of innovation

135

10.1 Background and batteries

Electric cars are not new. An electric car held the land speed record of 65.75 m.p.h. in 1899, but since the early prototypes, the various components of the systems have improved beyond recognition. The car industry is on the threshold of moving into electric traction on a large scale. General Motors intended at one time to have 25 per cent of its total production given over to electric cars by 1985. That target is perhaps a little optimistic. Nevertheless, it is estimated that by 1990 around seven million electric vehicles will be running on European highways (EEC Commission estimate). Similarly the United States' Department of Energy estimates that around eight million electric vehicles will be in use by the end of the century. But even these targets may not be realized.

Britain used to be considered the technological leader in electric vehicles. A large fund of experience is represented by the 44 000 milk floats running on British streets, and by the electric fork-lift trucks widely used in industry. Seventy per cent of such fork-lift trucks are electrically driven.

Figure 116 Lucas Chloride electric van

Major British companies such as Chloride and Lucas have invested heavily in electric vehicles. Lucas has concentrated its efforts on the adapted Bedford one-tonne van and has a fleet of sixty-five such vehicles with operating experience of 250 000 miles. In 1982 a joint company 'Lucas Chloride Electric Vehicle Systems Limited' was formed. This company plans to produce electric drive trains for four major manufacturers, to be used in variants of existing commercial vehicles. These are the Dodge 50 Series (Renault); Sherpa van (BL); Terrier truck (BL); Bedford CF van (GM) (Figure 116).

By 1986 Lucas Chloride plans to have built some 1500 electric vans and trucks. The initial costs are likely to be high, but the total lifetime costs are only 14 per cent higher than diesel equivalents (Figure 117). The first vehicle, the Dodge 50, will have a 'market support grant' from the DoI of £4000 for every vehicle. The batteries are to be improved lead–acid types, but both sodium–sulphur and nickel–zinc designs are being researched.

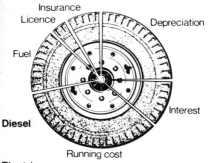

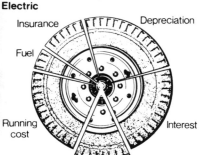

Figure 117 Comparative costs for diesel and electric commercial vehicles

There is a clear division between Britain and America in their strategies for electric vehicles. The United Kingdom is moving into the relatively unglamorous, low-performance delivery-vehicle market, while the United States designers are seeking a high-energy-density battery to power passenger cars.

The main inhibition upon the development of passenger cars is that a typical battery pack would be about one-third of the total vehicle weight, as opposed to a petrol engine, which is about one-sixth. The future growth of electric cars hinges about the development of lightweight batteries.

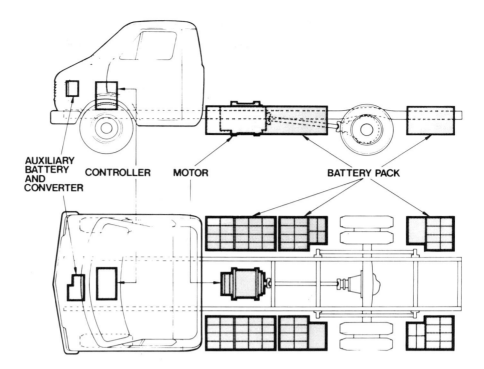

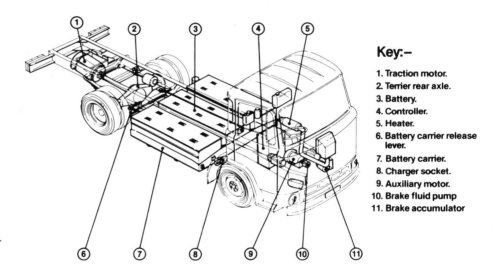

Figure 118 Dodge 50 electric van in production, 1982

AUXILIARY
BATTERY
AND
CONVERTER   CONTROLLER   MOTOR   BATTERY PACK

Key:–

1. Traction motor.
2. Terrier rear axle.
3. Battery.
4. Controller.
5. Heater.
6. Battery carrier release lever.
7. Battery carrier.
8. Charger socket.
9. Auxiliary motor.
10. Brake fluid pump
11. Brake accumulator

Figure 119 Prototype electric Terrier truck, launched by BL in 1982

Batteries

The battery systems currently on offer are varied and complicated. The three characteristics needed for an electric vehicle are:

high energy density;

quick, efficient recharging;

high number of recharging cycles before loss of efficiency.

Let's look at these characteristics in turn.

The energy density determines how much energy is available for a given weight of battery, which in turn determines the performance of the vehicle. The unavoidable fact is that batteries are heavy. A conventional lead–acid battery is about sixty times heavier than the amount of petrol to give equivalent energy. Even the best achievable energy densities by new battery forms are twenty times less than petrol. Design targets for battery systems are given in Table 16.

Table 16 Design targets for battery systems currently under development

Battery	Operating temp.	Open-circuit cell voltage/V	Capital cost/ ($ per kW h)	Cycle life	Energy density/ (W h per kg) theoretical	achievable	Out/in electrical efficiency
lead–acid	ambient	2.1	57	1000	175	50	75
nickel–iron	ambient	1.37	92	2000	267	60	55
zinc–air	ambient	1.65	92	600	1054	90	45
nickel–zinc	ambient	1.74	69	500	373	90	75
zinc–chlorine hydrate	ambient	2.12	41	500	461	150	65
lithium– titanium disulphide	ambient	2.5–2.0	46	1000	480	100	70–80
sodium– sulphur	300–500 °C	2.08–1.75	57	2000	758	130	75
lithium– iron sulphide	375–475 °C	2.1–1.6	57	1000	650	180	75

Source: *Ford Energy Report*, p.28.

The leading contenders for 'super batteries' are sodium–sulphur and zinc–chlorine. Other combinations such as nickel–zinc do not show much gain over improved lead–acid batteries.

The higher-energy-density batteries, such as sodium–sulphur, operate at high temperatures and use corrosive liquids and therefore require special precautions. The weight penalty in engineering safe and reliable batteries can bring the realizable energy density down as low as 110 W h/kg.

The major companies involved in electric vehicles have their own preferred battery systems. Gulf and Western (USA) have opted for a zinc–chlorine hydrate combination. This involves pumping the electrolyte solution through the system and chilling it to about 9 °C. This chilling inhibits the formation of dangerous chlorine gas. The result instead is a chlorine hydrate slush. These ancillaries, of course, generate further weight. General Motors have developed an electric vehicle powered by nickel–zinc batteries. In its early form this battery has a life span of only a few charges but has since been improved. (See Figure 120.)

The obstacles to the new batteries are in engineering and cost, rather than in basic chemistry. The development work so far has brought such systems into the near-term prospect from the long-term prospect. The basic materials are cheap, the energy potential is theoretically assured, but the new batteries have yet to be proven for reliability, safety, cost, lifespan and practical performance.

Recharging

The refuelling pause for a petrol-driven car is typically $3\frac{1}{2}$ min. Recharging an electric vehicle off the domestic supply would take about eight hours. Even if the technology of a 'quick' charge were developed, bringing charging time down to one hour, there would be an enormous current consumption at the charging station. Alternatively, the battery pack itself may be lifted out and replaced, swapping an empty battery for a full one, so to speak.

The number of recharging cycles the battery will sustain varies from system to system. The conventional lead–acid battery will take about 300 cycles although this can be improved to 700 cycles or more. A nickel–iron battery will run for 1500 cycles before replacement, but it gives off hydrogen when being charged and is generally poor in energy efficiency. Recharging is also a big problem for nickel–zinc. General Motors, who have vigorously promoted the nickel–zinc battery, extended their recharging life by vibrating them

Figure 120 Electric cars under development in the United States of America

(a)   General Motors Electrovette. A straightforward machine with a battery pack in the rear linked to an electric motor that drives the front wheels. Its nickel–zinc batteries take up half the space and weigh less than half as much as conventional lead–acid batteries of comparable power. The car carries its own charger, so it can be recharged at any 110 V outlet

(b)   Gulf and Western electric car. Chlorine, mixed with chilled water to form a slush of chlorine hydrate, is stored in the front. When the hydrate is warmed, it releases chlorine into the electrolyte, which is pumped back to zinc–coated plates in the cell stack. The chlorine reacts with the zinc to produce electricity and zinc chloride. In charging, the zinc is redeposited on the plates, and the chlorine pumped to the front and restored to hydrate form. The external charger also refrigerates the chlorine hydrate, which can be kept cool for four weeks in its insulated storage

(Source: Burck, 1980, p.78)

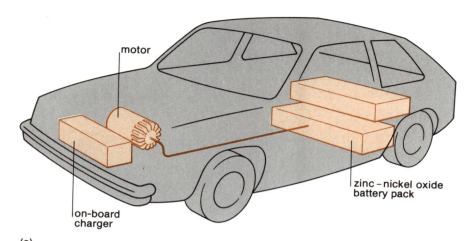

(a)

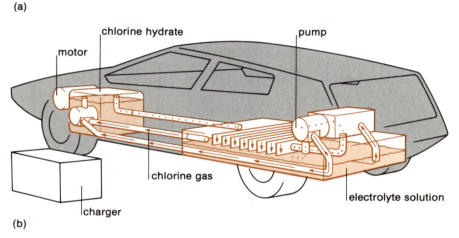

(b)

continuously. The zinc–chlorine system looks more promising. This battery has sustained over 1400 cycles, that is, the equivalent of 200 000 miles of motoring, way beyond the normal life of mechanical components.

What's more, the zinc–chlorine battery does not lose its rated power over 95 per cent of its discharge cycle. By comparison lead–acid batteries are down to nearly half their power by the time they are half-drained.

SAQ 35

What are the main battery types described here? List their main advantages or potential advantages.

10.2 Electric cars

What kind of cars can we expect from recent developments? The Gulf and Western car (Figure 120b) is an example of a promising prototype. This company has adapted a Volkswagen Golf (known as Rabbit in the United States), giving it a battery pack of 544 kg (1200 lb). The zinc–chlorine battery gives a range of 150 miles before recharging is necessary. This figure is based on an average speed of 55 m.p.h. and with a load of four passengers.

The Ford Motor Company is pursuing both lightweight lead–acid batteries and nickel–zinc batteries. For a battery pack of 500 kg, which would be realistic for a Fiesta-size car, Ford calculates the vehicle range for various systems is:

lead–acid	80 km,
nickel–zinc	140 km,
zinc–chlorine hydrate	240 km.

139

Figure 121 Electric Citystromers

Converted Volkswagen Colts being charged outside the GES offices in Essen

In the mean time improved lead–acid batteries seem likely to match the performance of petrol-driven cars.

A fleet of Volkswagen Colt cars have been equipped with heavy-duty lead–acid batteries by a German company, GES. The cars have a range of 40–50 km, at speeds of up to 96 km/h. The battery can be recharged by plugging it into one of a row of posts rather like parking meters (see Figure 121). However, the cost of the car, christened the Citystromer is £12 000 (1982 price). Mass-production would cut the price by more than a half, but the commercial success of the enterprise is very much dependent on the price of petrol.

Read the following extract from Walter Korff's book *Designing Tomorrow's Cars*.

An Electric Car for Tomorrow – The ETV-1

This is the first car to be built under the U.S. Department of Energy's Near Term Electric Vehicle Program. ETV-1 stands for ELECTRIC TEST VEHICLE-1. General Electric was the prime contractor and Chrysler Corp. was the major sub-contractor.

This vehicle seems appropriate for finalizing this book on designing tomorrow's cars. It is more aerodynamic, uses the latest chassis components, and is smaller and lighter than today's standard size cars. It is an advanced urban electric with improved performance that is amenable to mass-production by the mid-1980s. In quantities of 100,000 units the consumer price goal is $6,400 in 1979 dollars. The estimated life-cycle cost of the vehicle for its projected 10 year life is less than 18 cents per mile. This includes 2 cents per mile for scheduled maintenance.

[Figure 122 provides a three-quarter front view] of the sporty two-door hatchback for four people.

Figure [123] shows the key chassis features. These include a separately excited direct current motor which drives the front wheels, and the 18 high density lead-acid batteries located in a tunnel down the center of the vehicle. The Power Conditioning Unit (PCU) weighs 87 lb and provides the electronic controls for the armature and separate field control. The power transistor module in the chopper circuit weighs less than 1 lb and would reportedly cost less than $50 if produced in quantity. The microprocessor controls propulsion sequencing, "fuel" gauge computation and display, and programmed battery charging. A regenerative braking system recovers braking energy to recharge the propulsion batteries.

The transmission is a double reduction chain drive transaxle.

Figure 122 Electric test vehicle ETV-1 devised by General Electric and Chrysler

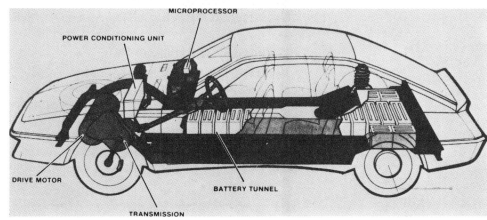

MICROPROCESSOR

POWER CONDITIONING UNIT

DRIVE MOTOR

TRANSMISSION

BATTERY TUNNEL

Figure 123 ETV-1 showing key features

GLOBE-UNION INC.

EV2-13 LEAD-ACID
ELECTRIC VEHICLE BATTERY

- 6 Volt
- 27.2 kg (60 lb)
- Unconventional, Computer-
  Designed Cell Geometry
- Left-Hand and Right-Hand
  Models

KEY

1. Thin, Lightweight, Durable
   Polypropylene Container and
   Cover Thermally Welded for
   a Leak-Free Assembly
2. Single-Point Watering
   System with Safety Venting
3. Low-Resistance, Through-
   the-Partition Intercell
   Welds
4. High-Efficiency, Computer-
   Designed Radial Grids
5. Optimized Active Materials
6. Submicro Polyethylene
   Envelope Separators with
   Glass Mat

Figure 124 Improved form of lead–acid battery

Figure [124] depicts the improved lead-acid battery. Note that the plates are oriented parallel to the long dimension of the case, rather than perpendicular. This battery has the same length and width as a golf-cart battery. Energy density of 17 W h/lb has been achieved and a cycle life of 500 discharge cycles is anticipated (70% discharge). Each battery weighs 60 lb. The battery system provides higher energy density than commercially available lead-acid batteries – an increase of about 30%.

General Specifications

Curb weight – 3320 lb – Gross weight 3920 lb.
Wheelbase 98 inches – Tread 55.5 inches front and rear.
Width 65.7 inches – Height 51.6 inches – Length 169.4 inches.
Ground clearance – 6 inches – Tire size P175/75 R 13.
Frontal Area – 19 sq. ft – C_d (weighted for yaw) 0.31.
Motor – Separately excited DC 20 hp continuous – 41 peak.
Voltage – 98 volts – Current rating 175 amps.
Speed range 0 to 5000 rpm – weight 214 lb. Efficiency 89%.
Size – 12 inch diameter by 20 inches long.

Power Conditioning Unit (PCU) – Transistorized chopper.
Armature control electronics – 0 to 2500 rpm.
Current rating – 200 amps continuous – 400 amps transient.
Frequency – 100 Hz to 2 Hz variable.

Field control/Battery charger electronics – 2500 to 5000 rpm.
On-board charging capacity – 110 volt/15 to 30 amps.
Off-board charging capacity – 220 volts/60 amps.
Recharge time – 10 hours at 30 amps.

Predicted Performance

Range – 35 mph contant – 122 miles.
   45 mph constant – 102 miles.
   SAE J227 driving cycle "D" – stop/start – 45 mph max. – 75 miles.
Acceleration – 0 to 30 mph – 8.9 seconds – 25 to 55 mph – 17.6 seconds.
Speed – Cruising 55 mph – Passing 60 mph.
Grade – Maximum – 17%.

(Korff, 1980, pp. 255–8.)

What are the main features of the performance of this prototype electric car?

The battery pack weighs 489.6 kg (1080 lb); each battery weighs 27.2 kg (60 lb). This gives a weight just less than one-third of the curb weight (curb, or kerb, weight means weight of the vehicle without occupants or luggage).

The cruising range is up to 122 miles at 35 m.p.h., but reduces to 75 miles on the suburban cycle of driving. This would reduce even further for the urban cycle of driving (Figure 125).

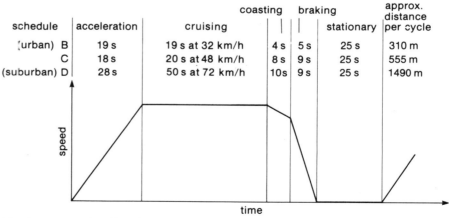

schedule	acceleration	cruising	coasting	braking	stationary	approx. distance per cycle
(urban) B	19 s	19 s at 32 km/h	4 s	5 s	25 s	310 m
C	18 s	20 s at 48 km/h	8 s	9 s	25 s	555 m
(suburban) D	28 s	50 s at 72 km/h	10s	9 s	25 s	1490 m

Figure 125 Urban driving cycle as specified for electric vehicles, SAE J277a

The battery recharging time is 10 hours at 30 amps and the anticipated cycle life is 500. Taking an optimistic view, the array of batteries would last for 37 500 miles, but for constant urban driving the battery life could drop to 7500 miles.

The ETV-1 described by Korff relies on conventional lead–acid batteries in an improved, lightweight form, but perhaps the large battery pack could be reduced if it were required only for steady cruising. The auxiliary power required could come from a smaller petrol engine.

10.3 Hybrid vehicles

Hybrid forms of motor car could be a near-term solution using combinations of existing technologies. If you think of the desirable characteristics in the performance of motor cars that we have mentioned already, you might arrive at a list such as this:

to optimize engine running and combustion conditions,

to have a better match between the engine and the wheels,

to be less dependent on oil-derived fuels,

to have a better engine-control system,

to recover kinetic energy on braking.

Now some of these requirements imply conventional petrol engines, but others imply electric power or alternative energy stores. This has led some designers to propose hybrid forms of car, where each major element is addressed to one performance phase.

At its simplest, a hybrid car has one element to cope with steady-state low-power driving, and another additional element to cope with intermittent surges of power needed for acceleration and hill climbing. Also, the efficient transition from one power source to the other is greatly assisted if an energy store can bank surplus energy. This could be achieved electrically in a battery, or mechanically in a flywheel.

The following extract makes the case for hybrid vehicles and goes on to draw general conclusions. The abbreviation NTHV stands for near-term hybrid vehicle, a programme of research initiated in the United States of America.

Read this article carefully and then attempt SAQs 36–9.

The IC engine/battery electric hybrid vehicle

The ic engine/battery electric hybrid utilizes both an internal combustion engine and electric motor either connected in the series arrangement, [Figure 126], or the parallel arrangement, [Figure 127]. In the series arrangement, the ic engine is generally sized to meet steady cruise and battery-charging requirements, with any excess power required for, say, vehicle acceleration coming from the batteries. However, since all energy supplied to the driven wheels passes through the electric traction motor, it (the motor) must be large enough to cope with the maximum road power and torque demand. The need for such an electric motor rated at maximum vehicle power plus a generator and batteries make this a physically heavy option and unsuited for the car and light commercial vehicle market. But, because of the electrical interconnection, the system does offer a high degree of flexibility in the positioning of power train components within the vehicle chassis, and does find application in the city bus sphere.

In the parallel arrangement both the internal combustion engine and electric motor provide power directly to the driven wheels with the ic engine again sized to meet the steady cruise requirement. The sizing of the electric motor now depends on the operating philosophy, but does not provide maximum road power. It can therefore be reduced in size. It may, for example, only provide accelerating power and cater for low speed performance as an 'electric vehicle'. For this reason the main effort in the smaller vehicle sector is towards the evaluation of this type of parallel hybrid configuration.

The development work on hybrids during the 1960s and early 1970s was primarily to reduce exhaust emissions; but the world oil crisis in the mid 1970s changed the emphasis such that more recent vehicle designs are aimed at either:
1 reducing the use of fuel from crude oil, or
2 minimizing the energy consumption for the user.

While both these aims are currently valid, from a government perspective the most pressing, in the medium term, is the former; while for the consumer it is the latter. To some extent both aims overlap with the more efficient ic engine vehicles [...] meeting both requirements to a certain degree. However, embedded in aim (1) is the so called 'petroleum substitution' option whereby a substantial proportion of energy demand is switched from oil-based fuels to electricity with its attendant wide fuel base. This is one of the declared aims of the NTHV Programme. Although such an aim can be partially met by penetration of the electric vehicle into certain parts of the road vehicle market, there are large areas where penetration will be difficult without a significant breakthrough in battery technology, and associated infrastructure, because of the range limitation of the electric vehicle. Table [17] shows a combination of the earlier data on energy used by type of vehicle, and second vehicle owners, and lists the vehicle group giving the largest energy saving, with particular consideration given to the car market. The sector with the most potential is that of the family car. However, [...] the concept of a range-limited vehicle will be difficult to 'sell', although limited penetration may be possible in the future with range-limited electric vehicles in the growing second car market, where the commuter car may find limited acceptance.

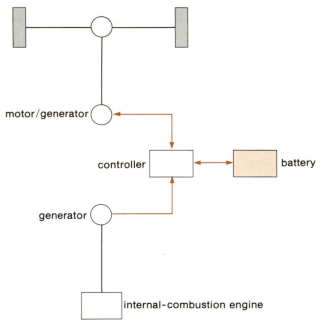

Figure 126 Components of a series internal-combustion engine/battery electric hybrid vehicle

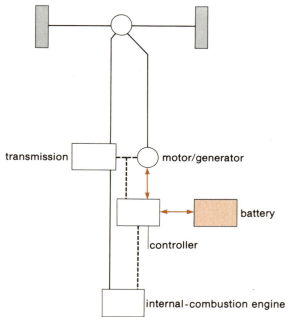

Figure 127 Components of a parallel internal-combustion engine/battery electric hybrid vehicle

Table 17 Energy used for road transport in 1980

Energy used			No of vehicles	
PJ		Type	10^6	
197	(45)*	Car 1200-1500	4.6	(1.0)
181	(41)	Car 1500-1800	3.5	(0.8)
123		Vans	1.2	
94	(21)	Car 1000-1200	2.2	(0.5)
72	(16)	Car 1800-2000	1.4	(0.3)
38		Bus	0.07	
12		Taxi	0.04	

* Bracketed figures refer to second car ownership.

Source: Department of Transport (1981), *Transport Statistics Great Britain 1970–80*, HMSO.

If an ic engine/battery electric hybrid car can be developed that meets either of the above objectives, and if it is not held to a strict range limit, as is the electric vehicle, the hybrid could have a significant impact in the future when, and if, there is a 'squeeze' on petroleum fuel. In particular, Table [17] suggests the initial development aim should be towards a family-sized car with a performance and specification similar to current models with an engine capacity of about [1500 c.c.].

Other development areas, but with lower energy savings and with more defined usage patterns for the ic engine/battery electric hybrid, are in the light delivery vehicle, city taxi and city bus. The market potential of such a hybrid vehicle in these areas could be limited, due to the possible impact of the electric vehicle, or electric vehicle with range extender. There is not likely to be a significant penetration into the heavy goods vehicle area because of its long-haul nature, and therefore this is most suited to ic or Joule (Brayton) engine.

The impact of hybridization in the city bus portion of the market is further complicated by the development of the Duo-bus concept, a vehicle using overhead power lines as one source. Such a concept could be attractive to countries that operate trolley-buses; but its impact in the UK is less easy to define because of the large capital cost of installing the overhead cables, even though restricted to the main city arteries.

Having identified the vehicles to be considered the duty cycle required for testing purposes must be specified. It is vital to make correct comparisons between hybrid vehicles and the equivalent internal combustion engine vehicle. If a large ic engine is replaced by an inadequate hybrid arrangement, and driven over a low demand cycle, the measured fuel savings of the hybrid are interesting, but not comparative. Unfortunately, it is easier to see when false comparisons are made and not so easy to see the correct criteria. For example, the top speed of an internal combustion engine is well over the UK legal maximum but the engine power is necessary for acceleration requirements at lower speeds. Perhaps an initial specification for a hybrid vehicle would include a specified top speed, possibly up a defined gradient; acceleration characteristics by specifying a time for 0-96 km/h, and an intermediate specification for overtaking purposes, as well as a minimum range requirement. In the first instance, such a specification should make the hybrid vehicle compatible with the conventional vehicle in traffic; although it can be argued that if significant fuel savings, and hence financial saving to the user, can be made, a reduced performance may be acceptable. However any such financial savings would need to be balanced with other cost factors – such as battery replacement cost, and any increased capital cost.

Although the above specifications relate to vehicle performance it is important that a hybrid vehicle is subjected to a representative comparison of performance and fuel economy with the conventional vehicle. The comparative conventional vehicle used should not be just a 'present day' model but should benefit from any developments that would be common place at a time consistent with that projected for the introduction of the hybrid.

In evaluating the fuel economy of the hybrid some representative test procedure is necessary. The current European method of determining fuel economy is to quote the fuel consumption during both a 'representative' driving cycle, [Figure 128], and at two constant speeds: 90 and 120 km/h. Although seemingly a straightforward test this procedure is complicated by the need for different cycles for different vehicle types; while even for one vehicle type different countries use cycles of different severity. In some cases cycles representative of different patterns are suitably weighted to obtain a net fuel efficiency. This difficulty of designing a cycle representative of actual vehicle use is discussed by Wouk [SAE 820269], but current cycles can be used to obtain a degree of performance comparison.

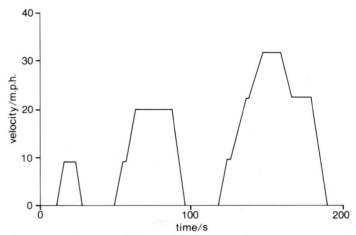

Figure 128 European ECE15 driving cycle

Testing an ic engine/battery electric hybrid adds a further complication. There are now two on-board fuels and a variety of control options so that the vehicle can, at least theoretically, operate with the electric motor or ic engine alone (parallel hybrid assumed), or as a hybrid. Consequently testing over a defined driving cycle, or at constant speed, in order to obtain fuel efficiency will depend on both the operating philosophy as well as the initial and final state of charge of the batteries. For example, fuel economy at constant speed could be a function of distance travelled, because initially, energy from the batteries could be used with the ic engine supplying a greater proportion of the power as distance increased, and battery state-of-charge reduced. Eventually, when a constant state of charge is reached, a value depending on the control design, a constant fuel economy will be recorded but this may well be at a distance in excess of the normal average distance travelled. The question then arises of suitably weighting the fuel economy figures for this one form of driving. To overcome such problems a number of test procedures for hybrid vehicles have been suggested, but they are substantially more complex than those currently used for conventional vehicle testing.

Although this brief discussion emphasizes the difficulty in testing a hybrid vehicle for comparison with the conventional, or other vehicle type, it is vital that a realistic comparison is made if the hybrid electric vehicle is to be a serious contender for future road transportation.

Conclusions

1 The transport sector has been taking an increasing share of UK primary energy, and in particular the petroleum products used in this sector, expressed as a ratio of the total delivered petroleum products, increased by 60% between 1973 and 1982. To a large extent this was due both to the ideal properties of petroleum as a fuel for transport and the massive manufacturing problems encountered in producing alternative prime movers.

2 Cars with engine sizes between 1000 and 2000 cc and light vans account for 40% of the energy used by the transport sector in the form of petroleum products, and this represents about 7% of the total UK primary energy requirement.

3 The concept of range limitation is not something at present considered by private or goods vehicle operators. Studies show that an 80 km daily range would satisfy 95% of car journeys and 67% of van journeys. Range limitation is a new concept, therefore to make a large market penetration with range-limited vehicles, without other external incentives, and/or government fiscal arrangements, the daily range would have to be 160 to 200 km.

4 In the near term electric vehicle battery options are the improved lead/acid, the nickel/zinc and the nickel/iron. The use of these batteries will lead to range improvements such that the electric vehicle could see a greater impact in the light commercial and city bus sector of the vehicle market in the future. However, because of range limita-

tions their impact in the car market is likely to be limited, particularly until the introduction of the long-term, high-performance sodium/sulphur and lithium/iron sulphide cells.

5 The ic engine/battery electric hybrid vehicle is most suited in terms of potential market sales, range, performance and savings in petroleum fuel to the car and light commercial vehicle sector. With the car fleet currently being replaced every 12 years, and the van fleet every 8 years, assuming no further growth this represents a potential market of over 1 M cars per annum and a van market of 150 000 pa. However, in order to switch to the hybrid from conventional engines would mean a large investment in new technology and manufacturing techniques.

6 In the car sector, analysis of the number of vehicles licensed and their energy use suggests that the major potential lies with an ic engine/battery electric hybrid car, with a performance and specification similar to current models having an engine capacity of about 1500 cc. Such vehicles currently account for about 40% of the energy used in the road transport sector.

7 To assess feasibility and optimum control strategy of an ic engine/battery electric hybrid vehicle, computer simulation (of the type of vehicle described) is necessary. Such work is currently being undertaken in the Engineering Department at Durham University.

(Bumby and Clarke, 1982, pp.8–11.)

SAQ 36

What are the main advantages of hybrid vehicles?

SAQ 37

Why do Bumby and Clarke think that a hybrid vehicle should have a performance and specification similar to a car with an internal-combustion engine of 1500 c.c.?

SAQ 38

What is the immediate difficulty in hybrid vehicle research, in the view of Bumby and Clarke?

SAQ 39

What are the contextual difficulties indicated in this article?

146

Figure 129 Ford hybrid test vehicle

Based on a Mustang, it has primary electric power and a 40 h.p. Wankel engine driving a generator in the rear. It allows a more flexible distribution of major mechanical components and can accommodate any kind of generator motor, including gas turbine or small high-speed diesel

Some of the disadvantages of hybrid vehicles remain similar to those of electric vehicles. Their range is limited (to around 200 km in the view of Bumby and Clarke) and they await the ever-expected breakthrough in battery technology. Further disadvantages of hybrid vehicles arise from the mix of technologies. For much of the time the vehicle is carrying a redundant system. The weight penalty may approach that of a full battery pack on a 'pure' electric car. In addition, hybrids depend on fine tuning and control. The conventional car engine is just about accessible to mechanics and owners for tinkering and adjustment. But a hybrid vehicle is carefully optimized and, unless there is good understanding of the underlying technical reasoning, tinkering and resetting can lose the advantages it offers.

10.4 Wider implications

It seems probable that by the end of the century there will be many thousands of electric vehicles on the roads of the United Kingdom. Some of the estimates of a few years ago look optimistic and overlook some of the technological difficulties of introducing a novel form of transport. Yet many manufacturers already are engaged in the production of electric cars. The 1980 World Car Catalogue lists forty-eight different electric designs, of which eleven are claimed to be in production.

Let's suppose for the sake of making conjectures that there would be a million electric vehicles in the United Kingdom by the year 2000, and that by that date about 10 per cent of car production would be given over to electric vehicles. Such a change, although relatively modest, would have widespread effects upon energy supply, and resources, upon the environment and motor manufacturing. Let's look at the implications.

Energy supply

If we suppose for the moment that one million electric vehicles are on the roads of the United Kingdom, what are the implications for energy supply? Firstly, there is plenty of spare capacity in the British electricity generating system. Secondly, most, if not all, of the electric vehicles could be recharged over-night at off-peak rates. One computer model indicates that up to three million extra electric vehicles could be accommodated without change to the supply system. An off-peak charge would be quite convenient if electric vehicles are used for short commuting trips, and you will recall that the bulk of car usage is over short distances. According to an American survey (*Ford Energy Report*, p. 30), 90 per cent of trips are return journeys of under forty miles (see Figure 130).

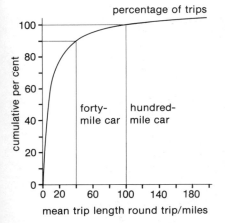

Figure 130 Trip length against vehicle range

It can be argued that vehicles that rely on electricity are inherently energy inefficient because of losses at the power station. In order to compare electric vehicles with conventional petrol-driven vehicles we need to look at the overall energy flow from primary source to propulsive work, that is, from the raw fuel to the wheels of the car.

Look at Figure 131. In these terms, the low efficiency of coal to electricity generation is compensated by the high efficiency in the electric vehicle itself. On the other hand the high efficiency of refining and distributing petrol is diminished by the low efficiency of conventional vehicles. With existing technologies, coal processed into a synthetic fuel and used to propel a heat engine gives the lowest efficiency of all (see Table 18).

Table 18 Comparative propulsion efficiencies

| Energy source | Percentage propulsion efficiency | |
	Heat engine	Electric
oil	13	18
coal	9	19

Source: Francis and Woollacott (1981).

In the future there are likely to be improvements in both the efficiency of electricity generation and in the internal-combustion engine. Accordingly, the most efficient pattern from raw fuel to propulsion could change significantly.

Resources

We have seen that some battery systems use cheap, widely available materials, such as sodium and sulphur, but others rely on less accessible metals. If our hypothetical one million electric vehicles were all propelled by lead–acid batteries then perhaps $\frac{1}{2}$ million tonnes of lead would be needed.

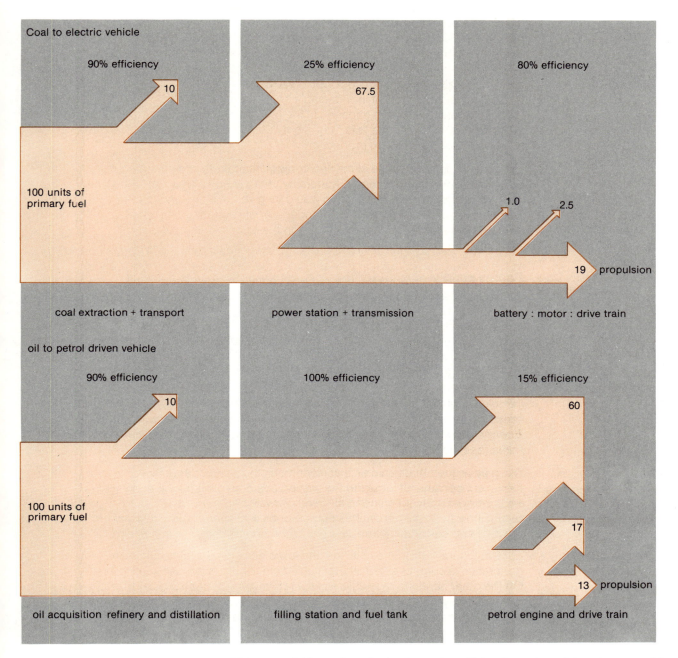

Coal to electric vehicle

90% efficiency   25% efficiency   80% efficiency

10   67.5

100 units of
primary fuel

1.0   2.5

19 > propulsion

coal extraction + transport | power station + transmission | battery : motor : drive train

oil to petrol driven vehicle

90% efficiency   100% efficiency   15% efficiency

10   60

100 units of
primary fuel

17

13 > propulsion

oil acquisition refinery and distillation | filling station and fuel tank | petrol engine and drive train

Figure 131 Propulsion efficiencies of electric and petrol-driven vehicles

Note: The efficiences for coal mining, crude-oil acquisition and refining, power-station generation and transmission, and synthetic fuel processes are generally accepted averages
(Source: Francis and Woollacott, 1981, Appendix 14, p.99)

This material is not lost, in that it can be efficiently retrieved, but there would certainly be an increase in the price of lead and hence in the cost of lead–acid batteries. Fortunately, the United Kingdom imports only about half (46 per cent in 1974–6) of the lead it uses. By comparison, the United States of America import only 4 per cent of their lead. The situation would be more difficult if zinc or nickel were used in the dominant battery type; the United Kingdom imports all its requirements of both metals.

All of which is to say that a complex network of trading and pricing make the final cost of new battery systems difficult to predict, and the commercial cost of batteries, as we have seen, is one handicap to electric vehicles.

Environmental benefits

Electric vehicles have certain strategic advantages. They are not tied into one fuel type, unlike petrol-driven cars. In the long term, when combustible substances are in short supply, it is conceivable that electric vehicles could be the dominant form of land transport. They can take advantage of the wide choice of methods of generating electricity (hydro, solar, nuclear, as well as coal burning).

149

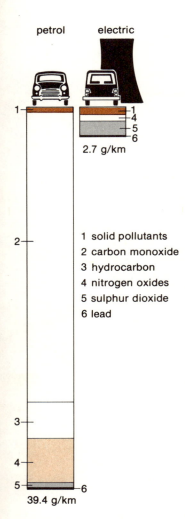

petrol | electric

2.7 g/km

1 solid pollutants
2 carbon monoxide
3 hydrocarbon
4 nitrogen oxides
5 sulphur dioxide
6 lead

39.4 g/km

Figure 132 Comparative pollution caused by internal-combustion engines and electric vehicles (Source: Aldous, 1978)

If there were a major switch over to electric vehicles, the level of urban pollution would drop significantly. Electric motors cause only about one-fifteenth the pollution of petrol engines.

However, electric cars would indirectly be responsible for more sulphur dioxide pollution from the chimneys of coal-fired power stations. Pollutants would be concentrated at the power stations, rather than in the urban areas where the vehicles are used. The control and dispersal of emissions are therefore more manageable and potentially less harmful. A further beneficial effect would be that car-generated noise levels would fall appreciably. Hybrid vehicles would come well within the legislated standards and pure electric vehicles are almost noiseless; in particular, there is no idling noise when the vehicles are stationary.

Performance and use

The electric car is intrinsically a short-range car. It could develop in its design as a purely urban vehicle, simpler and more manoeuvrable than a long-range petrol-driven car. In America, where two-car families are common, the electric car could become the typical second car for local trips.

However, the weight and bulk of the battery packs still present difficulties to be fully resolved. The internal-combustion engine and tankful of petrol set a hard standard to beat. The design of electric cars could be well packaged and aerodynamic, although with typical cruising speeds which are quite low (say 55 m.p.h.) some of the benefits of aerodynamics are lost.

If drivers could experience such new cars without the recent memory of speedy and responsive petrol-driven cars, then no doubt they would be impressed. But people's expectations of how cars will perform are very high, perhaps unnaturally high.

The only way an electric vehicle can approach the performance of a petrol-driven car is by reducing weight and compensating for the penalty of the battery payload. The impetus, then, is towards lighter cars without sacrificing safety standards. This corresponds with lines of development for conventional cars. Sometimes the two technologies coalesce into hybrid forms.

Manufacturing

In the mid-term, hybrid vehicles look quite promising, not only because of the potential gains in energy efficiency but because they represent an incremental step forward. Existing manufacturers of conventional cars powered by internal-combustion engines could more easily adapt their cars to be hybrid than change to totally new propulsive technologies.

A speaker at a conference on electric vehicles put the problem like this:

> . . .my problem in comparing electric with ICE vehicles is [not] whether we can get 125 million electric vehicles in the US. We cannot get 125 million of *anything* in the US in five or ten years. The crucial question is, when can we get *one* plant producing electric vehicles, or *one* alternative energy source, or *one* major non-conventional petroleum alternative fuel system rather than when can we get the large proportion of our transport system on to something other than petroleum.
>
> (Goodson, 1979, pp.45-6.)

If electric or hybrid cars are to be at all important in future transport, the next few years are crucial, not only in the development of lightweight batteries, but in the attitude of major manufacturers and Government. The year 2000 seems a long way off, but vehicles designed in the late 1980s and produced in the early 1990s will form the majority of vehicles on the road at the end of the century.

SAQ 40

What is the crucial technological limitation on existing electric vehicles?

SAQ 41

What would be the implications for electricity supply of one million electric vehicles being used on British roads?

SAQ 42

Which is the more efficient form of propulsion, an electric vehicle or a vehicle driven by an internal-combustion engine?

11
TUTOR-MARKED ASSIGNMENT
T263 05: THE FUTURE CAR

The assignment requires you to make some kind of conjecture about the future car. Yet your main impression from this block can only be that the car is technically extremely complex and subtle. More than that, no doubt, the later sections have created the further impression that the context of car design is fraught with difficulties: political, economic, social and environmental. In the assignment you should try to step back at first from the welter of detail and technical complexity and take a balanced view. The question to address is not only, 'What do you think *will* change?' but, 'What do you think *should* change?'

In your preparatory work for the assignment you should try to distinguish between short-term fluctuations and circumstances that will remain true over a long time span, between over-optimistic tactical thinking and more assured strategies. Similarly, you should be able to distinguish between inventive ideas and fully developed innovations. The enthusiasm of inventors is contagious. Their enthusiasm leads them into over-statements about the possibilities and effects of their inventions. The recent history of the car industry does not reveal many radical steps forward. Research and development are mainly directed towards energy efficiency and, although the impact on the future car will be dramatic, it is to be achieved by an accumulation of different means, by incremental innovations.

In the Supplementary Material for this assignment you will find a matrix that lays out the main technological changes covered in the block. It also includes some issues and circumstances that I have only briefly referred to, or hinted at. The matrix is laid out to show short-term, mid-term and long-term possibilities. It summarizes the block. You may like to fill in additional ideas from your own experience. Use it as a device for classifying which issues you would like to concentrate on for your assignment.

ANSWERS TO SELF-ASSESSMENT QUESTIONS

SAQ 1

The main arrangements of cylinders is shown in Figure 7. You should have included an in-line arrangement, a V-engine, a horizontally opposed engine and perhaps a radial engine.

SAQ 2

The main reason is that an internal-combustion engine gives *more power for less weight*. This is partly due to the efficiency of combustion, and partly due to the energy-to-weight ratio of the fuel. (Energy density of various fuels is discussed later in the block.)

SAQ 3

The mid 1890s saw the *clustering* of a number of patents into one new form of artefact. Some of these inventions derive from the bicycle and some were devised to meet specific technical problems of the motor car. Once the basic technology had crystallized into an automobile, the growth of the car industry took off very rapidly. See Figure 13.

SAQ 4

The essential features of mass-production are:

long production runs to justify the initial capital cost;

manufacture of standardized parts that can be interchanged;

division of work into short, specialized operations;

moving workpiece on an assembly line;

systematic planning and control of all the manufacturing operations for minimum material handling, and sequential flow of subassemblies and assemblies.

SAQ 5

Henry Ford, from the Model T onwards, produced cars cheaply and quickly. His methods of mass-production cut car production time down to $1\frac{1}{2}$ hours (about as long as you should take to study this section). Ford's ambition was not to produce a technologically advanced car (although this was true in his later career), but to manufacture a durable, simple design as cheaply as possible. The combination of cheapness and robustness meant that for almost twenty years the Model T was universally popular and accessible.

153

SAQ 6

The four main projects shown in the television programme 'Learning from the future' can be placed in our matrix like this:

Time ---→

Scope	Short term	Medium term	Long term
Features of cars	improved aerodynamics (Rupert Cambray)		
Types of car		car with more passengers (Kevin Rice)	electric car (Geoffrey Bird)
Alternative systems		hybrid car/motorbike (Mark Goodall)	

SAQ 7

The main problems of transmission are as follows:

The speed of rotation at the engine has to be geared down to the required speed of the wheels.

The coupling to the engine has to be intermittent to permit starting from rest and changing gear.

The torque of the engine has to be communicated efficiently to the wheels, usually through a 90° split at the differential.

The power required at the wheels varies according to driving conditions, hence a range of gears is needed to deal with moving from rest, going up hills, cruising at high speeds and so on.

The wheels turn at slightly different speeds when cornering. This difference has to be accommodated in the design of the transmission, usually through the differential.

SAQ 8

Here is my list of competing factors:

visibility for driver *v.* thick pillars for roll-over resistance,

sharp-edged creases for noise reduction *v.* rounded aerodynamic forms,

general structural rigidity *v.* controlled collapse,

local structural stiffness *v.* holes, channels for fixings,

ease of assembly *v.* complexity of body shell pressings,

and more generally

appearance and market appeal *v.* any of the functional requirements that have an effect on appearance.

SAQ 9

The three starting points are:

1   The experience and beliefs of senior management, which lead to a common-sense (or 'gut') feeling.

2   Feedback from their own sales organization about improvements, refinements or decline in popularity of some models and competitors' cars.

3   Market research conducted on potential customers and forecasting future trends.

SAQ 10

The main techniques described here and in section 4 of Unit 1 *An Introduction to Design* are:

sketches and renderings in perspective,

black-tape drawings at full size,

scale models, usually three-eighths scale,

full-scale models, usually of synthetic clay,

interior full-size mock-ups,

wind-tunnel tests of scale and full-size models,

dynamometer tests,

crash tests and computer simulations of crashes,

finite-element analysis for structure and weight,

road tests on different terrains and in different climates,

consumer clinics on prototypes.

SAQ 11

A computer is used to record and store three-dimensional information about body shapes.

A computerized system of finite-element analysis can examine structural rigidity of body shells and other components. The computer can simulate deformations and vibrations. Finite-element analysis leads to better structural and noise properties (in engines as well as body shells) and lighter, more efficient use of materials.

Computer programmes can simulate crash testing. The results can be compared with actual test crashes of prototypes. The number of destructive tests can therefore be reduced.

SAQ 12

The characteristics of a petrol engine that would give a performance of greater efficiency are:

less energy losses from cooling and exhaust;

open throttle, and therefore less air blockage;

weak mixture of fuel and air, called 'lean burn', but limited by tendency to misfire;

high compression, achieved with high-octane petrol, but limited by tendency to detonate.

SAQ 13

A diesel engine runs on a heavier fuel and relies on a high compression and fuel injection to make the fuel and air into an explosive mixture. It does not require a carburettor or a spark plug. The compression ratio for a diesel engine is around 20:1, whereas for a petrol engine the ratio is below 10:1.

Therefore a diesel engine is heavier, more sluggish and costlier than a petrol engine. However, in general it gives better fuel consumption and less pollution than a petrol engine.

SAQ 14

Apart from improved combustion, the main ways of conserving fuel are:

weight reduction,

air-drag reduction,

rolling resistance at tyres made less,

continuously variable transmission.

SAQ 15

Plastics components are lighter than steel equivalents, corrosion-free, and consume less energy in their manufacture.

SAQ 16

Paul Jaray initiated scientific testing of aerodynamic shapes. As chief designer of the Zeppelin Airship Works, he demonstrated the drag coefficient for a variety of simple shapes. More importantly, he showed the limiting drag coefficient of a passenger car to be of the order of 0.19.

Subsequent designs of aerodynamic cars can be viewed as manipulations and distortions of this teardrop form.

SAQ 17

Lay devised a series of interchangeable model components, from which a large variety of permutations were tested.

SAQ 18

The factors that contribute to the air resistance of a ground vehicle are:

frontal area presented to the air flow (usually approximates to maximum cross-sectional area);

velocity of vehicle (air resistance increases as the square of velocity);

density of the air (for precise results, tests are corrected to standard atmospheric values);

drag coefficient (the property of a given particular shape, see answer to SAQ 19).

You may also have mentioned other details of design that induce air drag, such as:

turbulence between the vehicle and the road, reduced by a front air dam;

turbulence in the wake, reduced by a rear spoiler;

local turbulence around the wheels and wheel arches;

local resistance caused by projections and discontinuities (wing mirrors, door handles, door pillars, etc.).

SAQ 19

The drag coefficient is the property of a given particular shape that indicates its ability to move easily through air (or any fluid). A low drag coefficient indicates greater slipperiness than a high drag coefficient. The limiting drag coefficient for a ground vehicle is around 0.12.

In car design, the drag coefficients of various vehicles are a shorthand way of comparing their aerodynamic performance. The comparative coefficients are deduced from wind-tunnel tests, in which all other properties (air velocity, frontal area, density) can be measured. Because the methods are empirical and not theoretical, the exact figures for drag coefficient are open to errors; for example, from scale reduction in the use of models and the interference characteristics of particular wind tunnels.

SAQ 20

The influences of the underbody and upper body on drag are roughly equivalent. Recent attention has been devoted mainly to upper body details, giving perhaps an 0.04 improvement in the drag coefficient. However, it is rarer to see underbody detailing refined, as for instance in the Ford Probe III. Such refinements could give a further improvement of 0.05 at the cost of increased weight.

SAQ 21

The drag coefficient for the Metro comes out fairly well when compared to its immediate rivals, the Renault 5, the Ford Fiesta, the Volkswagen Polo and the Fiat 127. Yet it compares less favourably with the Morris Minor and Austin A40, and less favourably still with radically novel cars, such as the Citroën DS19.

SAQ 22

We have seen that the ideal aerodynamic shape for ground vehicles is a kind of bisected teardrop. This form would give extra length to conventional cars, which, apart from causing difficulties in manoeuvring and parking, would carry a weight penalty. The car would need more power to carry the extra tail.

The low form of the ideal shape does not accommodate driver and passengers easily, unless they are in recumbent or semi-recumbent positions.

Similarly, the smooth shapes would require careful manufacture. This is perhaps less difficult in the double curvatures made by sheet steel presses, than in accommodating details such as flush windows, shallow door handles, and minimally projecting wing mirrors. Greater cost penalties are entailed in the details than in the overall shell.

The rounded front of such a vehicle would cause problems of driver visibility. Even with the recent generation of slant-nose cars there is difficulty in perceiving exactly the leading edge of the car. More importantly, the passenger space would have to adjust to the rounded contours. Future engineers and body designers will have to balance these costs against the benefits.

SAQ 23

There is a sense in which the extra fuel efficiency gained by aerodynamic shape can be thought of as available *free*. Cars have to have some shape, so that shape might as well be aerodynamic. The cost penalties between a body design that is aerodynamic and one that is not are relatively small. Designers do not have the option of *not* giving the car a body shape, in the way they do have options about *not* using diesel fuel, or turbochargers, or variable transmission.

SAQ 24

In the programme the problem of the Sierra wing mirror was discussed. Two factors had to be accommodated within aerodynamic design:

the protrusion allowed, and large field of view required, by law;

the collapsibility or 'foldability' of the mirror on contact (with pedestrians).

SAQ 25

Firstly, market research attempts to gauge public opinion in a perspective of six years, the time it takes to develop a car. Such cars in their basic form could last twenty years, like the Ford Cortina. The first dilemma is therefore one of predicting over long time scales, of making prophecies.

Secondly, huge investment programmes are necessary to launch a car – £700 million in the case of the Sierra – of which a small proportion is used for marketing and promotion (£2.5 million).

Thirdly, through marketing and salesmanship the company has to try to make the prophecy *self-fulfilling*.

SAQ 26

Normative forecasts
Technological conjectures, for example, new types of materials, engines and vehicles, voice-recognition computer controls.
Social goals, for example, public ownership of vehicles.
Personal values, for example, reference to 'noisy, evil-smelling' vehicles of today.

Predictive forecasts
Trend extrapolation, for example, rising labour costs of public transport (buses).
Evolutionary forecasting, for example, urban cars evolved to a simple cube.
Cross-impact analysis, for example, effects of legislation, growth of credit-card use.

SAQ 27

Figure 102 shows the energy split between rolling resistance and air drag as equal. Thus the speed would be about 50 m.p.h. for a conventional car with a drag coefficient of 0.45. Not only that, but a further proportion of energy is kinetic, therefore the car is still accelerating, and air drag would take progressively more energy.

SAQ 28

In general, most of the acute problems are being tackled. For example, improved engine burning, continuously variable transmission, turbocharging and aerodynamics are all well developed. Reducing the losses from cooling and braking is technically more difficult.

SAQ 29

In the low-speed gap Bouladon indicates a typical speed of about 7 m.p.h. over distances between 0.3 mile and 3 miles. See Figure 104.

In the high-speed gap he indicates a vehicle to travel at about 150 m.p.h. over distances between 30 miles and 300 miles.

So this first instance would be well fulfilled by a horse or bicycle, while the second instance would imply a high-speed train or a helicopter.

SAQ 30

The total post-war road deaths in the United Kingdom (1945–1980) are 209 244. The cost per death is estimated at £132 700.

Therefore the total cost is 209 244 × £132 700 ≈ £27 000 million.

SAQ 31

The most likely cause of the decline in casualties is the lower number of vehicle miles travelled as a consequence of the fuel crisis. The slower speeds adopted by motorists to conserve fuel may also have had a beneficial effect on accidents. Speed limits may also have had some effect.

SAQ 32

The survey of 1972 showed that the following features were thought to be disturbing:

noise and vibration,

dust and dirt,

fumes,

parking.

SAQ 33

Here is my ranking of new fuels based upon their state of development rather than efficiency of production and performance, or other considerations.

1   Sugar cane alchohol. Very important in Brazil, substituting for over 20 per cent of nation's fuel consumption.

2   Liquid petroleum gas. A petroleum by-product widely used in unconverted engines.

3   Corn ethanol. Used in the United States of America in a 1:9 mix with petrol (gasohol).

4   Methanol. Used experimentally in the United States of America, for instance, in a fleet of forty Ford Escorts.

5   Methane. Again, used in a few experimental vehicles, refrigerated into a liquid form.

6   Hydrogen. Still in early stage of development; most promising in a metal hydride form.

SAQ 34

This is my list of factors to weigh up:

Source.   Availability, cultivation difficulties; competition with other crops or uses; distribution of raw materials.

Processing.   Difficulty and cost of refinement; toxicity, corrosion; blending with petrol.

Delivered form.   Energy density; changes to engine design; necessary accessories (e.g. refrigeration, stainless steel).

Performance.   Octane rating; probable engine efficiencies; difficulties (e.g. vapour lock, carburation).

External.   Trade and tax base; price of oil; social difficulties; safety (fire, explosions, fumes).

You may arrive at a different list, but try to come to some feeling for the more important factors.

SAQ 35

The main battery types and their advantages are:

Lead–acid
Conventional technology well tried.
Can be made in lightweight form.

Nickel–zinc
Medium range possible (140 km).

Sodium–sulphur
Cheap basic materials.
High energy density (130 W h/kg).

Zinc–chlorine
High number of recharging cycles.
Long range possible (240 km).
High energy density (150 W h/kg).

Nickel–iron
High number of recharging cycles.

See Table 16 for further information.

SAQ 36

The hybrid vehicle, like the electric vehicle, can show improved fuel efficiency and reduced pollutant emission. Moreover, a hybrid vehicle can potentially deal with sustained surges of power in a way that is difficult for vehicles driven on batteries alone.

SAQ 37

The largest energy user in road transport in the United Kingdom is the medium-sized car, accounting for around 40 per cent of total energy used. A hybrid vehicle that could perform as well as a car in the middle of the 1200–1800 c.c. range would stand more chance of moving into this market.

SAQ 38

The testing of hybrid vehicles over a defined driving cycle is more complex than testing a car with an internal-combustion engine. Until the test procedures are specific about such things as the state of charge of the batteries and the control options between battery and engine, results are difficult to compare with conventional vehicle testing.

SAQ 39

The switch from conventional engines to hybrid forms requires large investment in new technology. Without subsidies or tax incentives or other forms of Government support, hybrid vehicles are unlikely to make a large penetration into the car market.

SAQ 40

Existing electric vehicles are limited by the range permitted by their batteries. This makes them suitable for commuting distances, say, round trips of 40 miles, but less suitable for long-distance driving. Researchers are pursuing more efficient, more powerful, lighter forms of battery.

SAQ 41

One million electric vehicles could be quite comfortably accommodated within the existing electricity supply system, particularly if they are charged at off-peak times.

SAQ 42

The electric vehicle is a more efficient machine than the conventional car with an internal-combustion engine, but great improvements are being made to the latter. If you take into account the fuel type, refining and generation and so on, then the propulsion efficiencies are quite similar:

coal to electric vehicles, 18 per cent;
oil to internal-combustion vehicles, 13 per cent.

See Figure 131.

CHECKLIST OF OBJECTIVES

Having studied all the related components of this block you should be able to do the following:

Section 1

1   Give an outline of the early origins of motor cars.

2   Explain the principles of an internal-combustion engine.

3   Describe the main cluster of inventions that led to the motor car.

4   Describe the essential features of mass-production.

Section 2

5   Explain the main problems involved in the transmission system of cars.

6   List the main factors involved in body-shell design.

7   Exemplify conflicting factors in design as revealed by car body design.

Section 3

8   Give an account of some of the ergonomic problems of using motor cars.

9   Make diagrammatic sketches of car interior layouts.

10   Make an anthropometric analysis of car packaging using the *Humanscale* charts.

Section 4

11   Describe the factors that initiate the process of design in the car industry.

12   Give illustrations of the main design techniques used in the development of cars.

13   Exemplify the usefulness of computer-aided design in the car industry.

Section 5

14   Describe the main ways to improve the fuel efficiency of the internal-combustion engine.

15   Distinguish between the characteristics of a petrol engine and a diesel engine.

16   Describe new systems of continuously variable transmission.

17   Explain the main improvements in fuel efficiency that can be achieved with conventional cars.

Section 6

18   Outline the early history of aerodynamic research.

19   Explain the general principles of aerodynamics.

20   Exemplify the conflict between aerodynamics and other factors in car design.

Section 7

21   Explain the difference between normative and predictive forecasts of the future, and give some examples.

22   Explain what is meant, in futures forecasting, by a 'scenario'.

23   Identify and give examples of the use of some basic techniques of futures forecasting.

Section 8

24   Weigh up the benefits and defects of the motor car as a system of transport.

25   Describe the main areas of mechanical inefficiency in the motor car.

26   Explain what is meant by the term 'transport gaps'.

27 Give an account of the main financial costs of private motoring.

28 List the main hazards of motor traffic.

29 Estimate the costs of road accidents in the United Kingdom.

Section 9

30 Give a brief account of the energy context of car design.

31 Distinguish between various types of new fuels.

32 Describe the state of commercial development of some alternative fuels.

33 Discuss the implications of a major change to plant-derived fuels.

Section 10

34 Describe recent developments in electric and electric hybrid cars.

35 Explain the main requirements for batteries appropriate to electric vehicles.

36 Compare the efficiencies of electric vehicles and vehices powered by internal-combustion engines.

37 Give an account of the implications of the widespread use of electric vehicles.

REFERENCES

A. Aldous (1978), 'The electric way ahead for road transport', *Electronics and Power*, vol. 24, no. 4, April, pp.289–93.

W. Bernhardt (n.d.), 'Alternative fuels from biomass and their use in transport', Volkswagen.

K. Bhaskar (1979), *The Future of the UK Motor Industry*, Kogan Page.

G. Bouladon (1967), 'The transport "gaps"', *Science Journal*, vol. 3, no. 4, April, pp.41–6.

C. Buchanan (1964), *Traffic in Towns*, Penguin.

J.F. Bumby and P.H. Clarke (1982), 'The role of the internal-combustion engine/battery electric hybrid vehicle in the UK', *Energy World*, December, no. 98, pp.2–11.

C.G. Burck (1980), 'Can this car electrify America?', *Fortune*, vol. 102, no. 1, 14 July, pp.77–80.

D. Burgess Wise (1982), *Ford USA Pocket History*, Automobilia, Milan.

G. Bylinski (1979), 'Biomass: the self-replacing energy resource', *Fortune*, vol. 100, no. 6, 24 September, pp.78–81.

E. Callenbach (1975), *Ecotopia*, Banyan Tree Books, Berkeley, California (Pluto Press, 1978).

C.F. Caunter (1970), *The Light Car*, HMSO.

Central Policy Review Staff (1974), *The Future of the British Car Industry*, HMSO.

A.C. Clarke (1974), *Profiles of the Future* (2nd edn), Gollancz.

D.L. Clarke (1978), *Analytical Archaelogy* (2nd edn, rev. R. Chapman), Methuen.

T. Curtis (1981), 'Low-loss motor car – no owners yet', *New Scientist*, vol. 90, no. 1255, 28 May, pp.546–9.

D. Downs (1978), 'The socially acceptable power plant', *Proceedings of the Institution of Mechanical Engineers*, vol 192, December, pp.343–57.

S. Encel, P. Marstrand and W. Page (eds.) (1975), *The Art of Anticipation*, Martin Robertson.

J. Fairbrother (1981), *Fundamentals of Vehicle Bodywork*, Hutchinson.

Ford Energy Report (1982), Special publication SP1, International Association for Vehicle Design.

R.J. Francis and P.J. Woollacott (1981), *Prospects for Improved Fuel Economy*, HMSO.

C. Freeman and M. Jahoda (eds.) (1978), *World Futures*, Martin Robertson.

R. Fry (1980), *The VW Beetle*, David & Charles.

E. Fuhrmann (1979), 'Creation of the Porsche 928', *International Journal of Vehicle Design*, vol. 1, no. 1, pp.75–84.

R. E. Goodson (1979), discussion comment in Electric Vehicle Development Group Third International Conference, *Resources for Electric Vehicles and their Infrastructure*, Peter Peregrinus, pp.45–6.

C.L. Gray and F. von Hippel (1981), 'The fuel economy of light vehicles', *Scientific American*, vol. 244, no. 5, May, pp.36–47.

E. Jantsch (1967), *Technological Forecasting in Perspective*, OECD.

J.C. Jones (1980), *Design Methods* (2nd edn), Wiley.

W.H. Korff (1980), *Designing Tomorrow's Cars*, M-C Publications.

K. Ludwigsen (1970), *The Time Tunnel*, Society of Automotive Engineers.

L.T.C. Rolt (1956), *A Picture History of Motoring*, Hulton Press.

F. Sands and V. Batty (1974), 'Road traffic and the environment', in M. Nissel (ed.) *Social Trends no. 5 1974*, HMSO.

F.K. Schenkel (1969), 'A review of the state of the art of automobile aerodynamic research', Cornell Aeronautical Laboratory.

M.W. Thring (1973), *Man, Machines and Tomorrow*, Routledge.

A. Toffler (1971), *Futureshock*, Pan.

C.H. Waddington (1977), *Tools for Thought*, Cape.

C.H. Waddington (1978), *The Man-made Future*, Croom Helm.

M.E. Ware (1976), *Making of the Motor Car 1895–1930*, Moorland Publishing.

G. Willis (1972), *Technological Forecasting*, Penguin.

Acknowledgements

Grateful acknowledgement is made to the following people and organizations for permission to reproduce material in this block:

Text

pp.100–101 Ernest Callenbach, Pluto Press and Basic Books Inc. for extracts from E. Callenbach (1975), *Ecotopia*, Banyan Tree Books/Pluto Press, © 1975, 1978 Ernest Callenbach; p.106 the estate of C.H. Waddington, Jonathan Cape and Basic Books Inc. for extract from C.H. Waddington (1977), *Tools for Thought*, Cape/Basic Books; pp.107–9 John Wiley and Sons Ltd for extract from J.C. Jones (1980) *Design Methods* (2nd edn), Wiley, © 1981 John Wiley; pp.140–42 W.H. Korff (1980), *Designing Tomorrow's Cars*, M-C Publications, Burbank, California; pp.144–6 J.R. Bumby, P.H. Clarke and the Institute of Fuel for excerpt from J.R. Bumby and P.H. Clarke (1982), 'The role of the internal combustion engine/battery hybrid vehicle in the UK', *Energy World*, December, no. 98, pp.2–11.

Tables

Table 2 J. Fairbrother (1981), *Fundamentals of Vehicle Bodywork*, Hutchinson; Table 4 W.H. Korff (*op. cit.*); Tables 5, 6, 14 and 16 Ford Motor Company Limited from *Ford Energy Report* (1981); Table 9 Consumers Association from *Motoring Which?*, April 1982; Table 18 reproduced by permission of the Controller of HMSO from R.J. Francis and P.N. Woollacott (1981), *Prospect for Improved Fuel Economy and Fuel Flexibility in Road Vehicles*, Energy Paper no. 45.

Figures

Figure 1 Terence Cuneo, photograph supplied by the Science Museum, London; Figures 2, 16, 23 and 74 National Motor Museum, Beaulieu; Figures 3, 17, 18, 48–53, 55–7, 60, 85, 86, 96(right), 113 and 129 Ford Motor Company Limited; Figures 4, 9, 10, 24(b) and 73 Peter Roberts Collection; Figures 6, 45 and 78 Volkswagenwerk AG; Figure 11 BBC Hulton Picture Library; Figure 12 © Hennerwood Publications Limited, from M.A. Smith (1979), *The Car*, Sundial Publications/Cathay Books; Figure 14 John Hillelson Agency Limited, photograph: L'Illustration/Sygma; Figures 15 and 21(a) Crown Copyright, Science Museum, London; Figure 24(a) *Motor*; Figure 27 L.T.C. Rolt (1956), *A Pictorial History of Motoring*, Hulton Books; Figure 28 Hutchinson, from Fairbrother (*op. cit.*); Figure 35 Car Styling and Bertone; Figure 43 Mercedes–Benz (United Kingdom) Limited; Figure 44 Vauxhall Motors Limited; Figure 54 BMW (GB) Limited; Figure 58 Motor Industry Research Association; Figure 59 United Press International (UK) Limited; Figure 61 Renault UK Limited; Figure 65 © Institution of Mechanical Engineers from D. Downs (1978), 'The socially acceptable power plant', *Proceedings of the Institution of Mechanical Engineers,* vol. 192, pp.343–57; Figure 67 Automotive Products plc, Leamington Spa; Figures 68(a), 69, 83, 106(a) and 122–4 W.H. Korff (*op. cit.*); Figure 71 Citroën Cars Limited; Figure 76 Arwin/Calspan Advanced Technology Center, from F.K. Schenkel (1969), 'A review of the state of the art of automobile aerodynamic research', Cornell Aeronautical Laboratory Inc; Figure 93 John Hillelson Agency Limited, photograph: Regis Bossu/Sygma; Figures 14 and 95(top) General Motors Corporation; Figure 95(bottom) Chrysler Corporation; Figure 96(left) *Guardian*; Figure 97 Professor M.W. Thring; Figures 104 and 105 *New Scientist*, from G. Bouladon (1967) 'The transport gaps', *Science Journal*, vol. 3, no. 4, pp.41–6; Figure 109 reproduced by permission of the Controller of HMSO from *Road Accidents in Great Britain* 1980; Figures 116 and 118 Lucas Chloride EV Systems Limited; Figure 117 Morgan-Grampian (Publishers) Limited, from V. Wyman (1982) 'Electric vehicles – the positive approach', *Engineer*, vol. 255, no. 6608; Figure 119 © IPC Business Press Limited, from *Electrical Review*, vol. 210, no. 17, 30 April 1982; Figure 120 © Time Inc., from C. Burck (1980), 'Can this car electrify America?', *Fortune*, vol. 102, no. 1, pp.77–80; Figure 121 Gesellschaft für elektrischen Strassenverkehr mbH; Figure 131 reproduced by permission of the Controller of HMSO from Francis and Woollacott (*op. cit.*).

The *Design: Processes and Products* Course Team

Chairman
David Walker

Authors
Godfrey Boyle
Stephen Brown
Nigel Cross
Robin Roy
Brenda Vale
Robert Vale

Reading members
Sheila Cameron
David Elliott
Alex Godden
Peter Smith

External assessor
Ken Baynes (Royal College of Art)

Consultants
Tony Curtis (Editor, *Motor* magazine)
Paul Gardiner (University of Sussex)

BBC
Phil Ashby
Kevin Newport
Colin Robinson
Bill Young

Course manager
Ernest Taylor

Editor
Robin Kyd

Librarian
Caryl Hunter-Brown

Graphic artist
Roy Lawrance

Graphic designer
Rob Williams

Secretaries
Kitty Gleadell (Course Secretary)
Olive Ainger

Design: Processes and Products